T0237286

Fundamentals of Propulsion

V. Babu

Fundamentals of Propulsion

Ane Books
Pvt. Ltd.

V. Babu
Department of Mechanical Engineering
Indian Institute of Technology Madras
Chennai, Tamil Nadu, India

ISBN 978-3-030-79947-2   ISBN 978-3-030-79945-8   (eBook)
https://doi.org/10.1007/978-3-030-79945-8

Jointly published with ANE Books India
In addition to this printed edition, there is a local printed edition of this work available via Ane Books in
South Asia (India, Pakistan, Sri Lanka, Bangladesh, Nepal and Bhutan) and Africa (all countries in the
African subcontinent).
ISBN of the Co-Publisher's edition: 978-81-80522-52-9

This Springer imprint is published by the registered company Springer Nature Switzerland AG
The registered company address is: Gewerbestrasse 11, 6330 Cham, Switzerland

*Dedicated to my wife Chitra and son
Aravindh for their enduring patience and love*

Preface

This book is a sequel to my earlier book "Fundamentals of Gas Dynamics". The material in this book has grown out of my class notes for the postgraduate course on Air Breathing Engines that I teach at IIT Madras. These classes had a heterogeneous mix of B.Tech., M.Tech., M.S. and Ph.D. students. As a result, I did not assume too much about the knowledge of the students on the topics discussed in the class and I usually spent a considerable amount of time teaching the fundamental concepts. I have adopted the same practice in writing this book also. Chapters 2–4 in the book are devoted to the discussion and development of important concepts in the areas of gas dynamics, turbomachines and combustion, which, in my opinion are essential for learning propulsion. Theoretical and practical design aspects of turbojet and turbofan engines are discussed in Chaps. 5 and 6. Thermodynamics and thrust calculation are discussed separately in Chap. 7. I have always felt that this separation allowed the students to grasp the concept and ideas better without getting bogged down in the details of calculations. As with my earlier book, my intention was not to write a book that was exam-oriented but one that will help students improve their understanding of the subject matter. The examples and exercise problems are drawn from practical applications to enable the student to appreciate the subtleties in the use of the concepts.

One striking feature of the book, perhaps, is the complete lack of any pictures of actual aircraft engines. Although many such pictures are available in the internet, due to commercial copyright restrictions, I have not been able use them in the book. In this connection, I wish to thank Rolls-Royce plc for generously permitting me to use a lot of images available in their corporate web site. The internet is a rich source of many interesting and sometimes rare pictures of aircraft engines and related material. I urge the readers to take advantage of this and broaden their knowledge.

I have suggested a list of books at the end (in alphabetical order) for the students to consult. These are classics (except the first one!) which the student should hopefully be able to appreciate after reading this book. The readers should realize that the concepts and ideas discussed here are not new, but have merely been presented in a different manner arising from my experience in teaching them. If there are any errors or if you have any suggestions for improving the exposition of any topic, please feel free to communicate them to me via e-mail (vbabu@iitm.ac.in).

I am very fortunate in that I have been blessed with good teachers and good students. This has enabled me to sharpen my understanding of the fundamental concepts in the subjects that I teach as well as my areas of research. I would like to once again thank my teachers Sri. S. Sundaresan, Profs. R. Bodonyi, M. Foster and T. Scheick, who, as teachers, inspired me to a great extent. One person to whom I owe the most is Prof. S. Korpela of the Ohio State University. I learnt from him not only how to be a good academic researcher, but, more importantly, how to be a good human being as well.

I am grateful to my senior colleague Prof. K. Ramamurthi for taking the time and effort not only for reviewing almost all the chapters in the book, but for writing additional material as well. I have benefited tremendously from his comments and suggestions. I am greatly indebted to him.

I wish to express my heartfelt gratitude to my parents who endeavored so much to give me a good education. They have given a lot to me but received very little in return.

Finally, I would like to thank my students S. Maathangi, P. S. Tide and K. Kumaran for diligently working out the examples and exercise problems and proof reading the manuscript.

Chennai, India   V. Babu
October 2020

Contents

About the Author

Dr. V. Babu is currently a Professor in the Department of Mechanical Engineering at IIT Madras. He received his B.E. in Mechanical Engineering from REC Trichy in 1985, M.S. and Ph.D. from the Ohio State University in 1989 and 1991. He worked as a Post-Doctoral researcher at the Ohio State University from 1991 to 1995. He was a Technical Specialist at the Ford Scientific Research Lab, Dearborn, Michigan from 1995 to 1998. He joined IIT Madras at the end of 1998. He received the Henry Ford Technology Award in 1998 for the development and deployment of a virtual wind tunnel. He has four U.S. patents to his credit. He has published technical papers on simulations of fluid flows including plasmas and non-equilibrium flows, computational aerodynamics and aeroacoustics, scientific computing and ramjet, scramjet engines. His primary research specialization is CFD and he is currently involved in the simulation of high speed reacting flows, prediction of jet noise, simulation of fluid flows using the lattice Boltzmann method and high performance computing.

Chapter 1
Introduction

Interest in propulsion systems for aircraft started with the first powered flight by the Wright brothers. Today, interest in aircraft propulsion systems is driven predominantly by the fact that it is a multi-billion dollar industry, serving both commercial and military applications. In the commercial sector, fuel economy, emissions, noise, safety and ease of maintenance are the primary issues. These issues are important in military applications also, although to a lesser extent. The primary issues are flight speed (usually in the high subsonic, supersonic range), thrust, performance under off-design conditions (important for maneuverability) and acceleration capability (important for combat and short takeoff, landing). In this book, we focus primarily on aircraft engines for subsonic flight speeds. In the last chapter, some propulsion systems used for supersonic flight speeds are discussed briefly, to give an idea of the issues involved in the design of such systems.

1.1 Thrust

Thrust is generated in almost all propulsion systems (land, water or air), on the basis of Newton's third law. In the latter two cases, the thrust force is generated by taking a certain mass flow rate of fluid and imparting a velocity change to it in the propulsion device. In land-based systems, forward force on the vehicle is generated due to the reaction force exerted by the ground on the vehicle (equal and opposite to the traction force exerted by the rotating wheels of the vehicle on the ground). Thus, it can be seen that in this case, the coefficient of friction between the wheels and the ground plays a major role in the generation of the motive force, independent of the engine. This is in contrast to the other two cases where the thrust generated is dependent only upon the engine. The thrust is, in fact, equal to the product of the mass flow rate of the fluid and the velocity change imparted to it, viz.,

© The Author(s), under exclusive license to Springer Nature Switzerland AG 2022
V. Babu, *Fundamentals of Propulsion*,
https://doi.org/10.1007/978-3-030-79945-8_1

$$T = \dot{m}(V_e - V_{\text{vehicle}}).$$ (1.1)

Here, T is the thrust, $\dot{m}$ is the mass flow rate of the fluid[1] undergoing momentum change, V_e is the velocity of the fluid as it leaves the propulsion device and V_{vehicle} is the forward speed of the vehicle. Equation 1.1 is of fundamental importance in propulsion.

In the case of a rocket, Eq. 1.1 simplifies to

$$T = \dot{m}V_e$$ (1.2)

since the rocket does not take in air but carries the required oxidizer on-board. At first sight, it appears from Eq. 1.2 that the thrust for a given value of $\dot{m}$ is considerably higher for a rocket than an aircraft engine. It must, however, be borne in mind that a bulk of the thrust in the case of the former is used for lifting the propellants (fuel + oxidizer) itself. Hence, an appropriate metric for comparing the performance of different aircraft engines is $T/\dot{m}$, while for rocket engines, it is $T/\dot{m}g$, where g is the acceleration due to gravity. The former quantity with units m/s is called specific thrust and the latter with unit s is called specific impulse.

1.2 Modes of Thrust Generation

Before discussing different modes of thrust generation, it is instructive to see what is required of a propulsion device. An examination of Eq. 1.1 reveals that a propulsion device should

capture a certain amount of mass flow rate of air: At first sight, it appears as though larger mass flow rates are better since the thrust produced will be more. However, at any given altitude, the captured mass flow rate of air is directly proportional to the flight speed V_{vehicle} and the frontal area of the propulsion device. But the aerodynamic drag is proportional to the frontal area and the square of the flight speed. Hence, the higher frontal area places a limitation on the flight speed.

impart a certain change in the velocity of the air: Note that this change can be a change in the direction, magnitude or both. A change in the magnitude of the velocity (in this case, an increase) has to be accomplished by the addition of work or heat. In the latter case, the addition of heat results in a significant increase in the temperature of the working fluid. Consequently, there will be practical limitations on the amount of heat that can be added (and hence the velocity increase that can be achieved) from metallurgical considerations. Limitations on the amount of

---

[1] Many of the concepts that we discuss can be applied whether the working fluid is water or air. Since our primary interest is in aircraft propulsion, we will assume the working fluid to be air from here onwards.

work that can be added arise from weight considerations since the size of the powerplant increases with the increasing amount of work addition.

It is clear from the above discussion that a practical propulsion device can generate thrust by either having the first term ($\dot{m}$) as large as possible with the second term being small or *vice versa*. A propeller achieves the former, while a jet engine achieves the latter.

1.2.1 Propellers

Since a propeller moves a large mass of air, the diameter of the blade and hence the frontal area is high. Also, the tip of the propeller blade rotates at a higher speed than the root. Even for low subsonic flight speeds, tip speeds can easily approach or exceed the speed of sound. This results in shock waves near the tips, which can drastically decrease the aerodynamic efficiency of the propeller, while increasing aerodynamically generated noise. Thus, propeller-driven aircraft are limited to low subsonic flight speeds. On the other hand, the second term in Eq. 1.1 is small for a propeller and so heat addition is not required. Since the propeller handles only cold air, material and metallurgical considerations arising from the handling of high-temperature gases do not arise. Also, since the velocity change is small, aerodynamically generated noise is quite low. It is for these reasons that the propulsive efficiency of a propeller is quite high at low subsonic flight speeds.

1.2.2 Jet Engines

In a jet engine, the captured air is compressed to high pressure followed by heat addition (in a combustor) to increase its temperature. This increases the enthalpy of the air, and after extraction of the work (in a turbine) required for compression, it is sent to a nozzle. Here, the remaining enthalpy of the air is converted to kinetic energy and it leaves the nozzle at a high exit velocity, V_e. The compression, heat addition and expansion in the turbine are accomplished in a gas turbine and the thrust generation in the nozzle. Thus, the gas turbine acts as a gas generator generating a stream of high enthalpy gas. Since the jet engine has to work at high pressures and temperatures, structural issues related to materials and metallurgy are critical. Also, since the exit velocity of the air (hot gas) is quite high, the mixing of this high-speed stream with the surrounding air results in high levels of aerodynamically generated noise. However, on the positive side, flight speeds in the high subsonic–supersonic range are possible with the jet engine.

Figure 1.1 gives an illustration of the propulsion technologies for various flight Mach numbers. The turboprop and turbofan engines are widely used today in commercial passenger aircrafts. The basic turbojet engine, although still in service, is

Fig. 1.1 Illustrative view of propulsion technologies for different flight Mach numbers

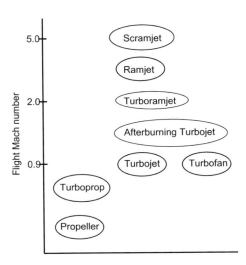

being replaced by turbofan engines worldwide. Turboprops are used primarily for short-haul passenger operations or for carrying cargo.[2] Afterburning turbojets, turbo-ramjet and ramjets are used exclusively in military aircrafts. The notable exception to this is the Concorde, which is the only civilian aircraft with an afterburning turbo-ramjet engine. The scramjet has a hypersonic propulsion technology that is still in the development stage. In the following chapters, we focus our attention on turbo-jet, afterburning turbojet and turbofan engines. In the last chapter, we discuss the turboramjet, ramjet and scramjet engines briefly.

1.3   Types of Aircraft Engines

Before and during World War II, propellers were used exclusively in aircraft propulsion. The propellers were driven by reciprocating piston engines, with the cylinders arranged in-line or radially. After World War II, the increasing need for higher flight speeds for both commercial and military applications signaled a shift to the use of an alternative powerplant (gas turbine) and mode of thrust generation (jet). Figure 1.2 shows a propeller driven by an in-line piston engine and a gas turbine.

In aircraft propulsion application, an important metric for assessing the relative performance of different engines is the power-to-weight ratio. If Engine A produces the same power as Engine B, but weighs less, then obviously Engine A is better. From Fig. 1.2, it can be seen that the working fluid undergoes the same sequence of

---

[2] Bombardier Aerospace of Canada and ATR of France are the leading manufacturers of turboprop aircraft today, with engines from Pratt and Whitney. Turbofan engines are manufactured by Rolls-Royce, General Electric and Pratt and Whitney and these are used in aircrafts manufactured by Boeing, Airbus and Embraer.

Fig. 1.2 Illustration of piston engine and a turboprop engine. *Courtesy* Rolls-Royce plc

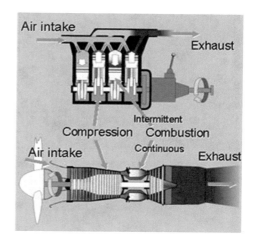

processes, namely, suction, compression, expansion and exhaust, in both the piston engine and the gas turbine. However, since each cylinder delivers power only once in four strokes, this sequence is intermittent in the former, while it is continuous in the latter. Consequently, the power-to-weight ratio of a gas turbine engine can be superior to a comparable piston engine by as much as a factor of 10. Reducing the number of stokes per cycle (i.e., from 4 to 2) and increasing the number of cylinders are possible strategies for improving the power-to-weight ratio of a piston engine. The first option increases the power delivered per revolution without any increase in the weight. The second option also achieves the same end effect but with a substantial increase in the weight due to the increased number of cylinders. For this reason, gas turbines have completely replaced piston engines in aviation applications.

Today, gas turbines are used in propulsion and non-propulsion applications. Figure 1.3 shows some of these applications. In the turboprop engine, as mentioned earlier, the propeller is driven by the gas turbine and generates the thrust. The gas turbine acts as a gas generator and the high enthalpy exhaust gas from the gas turbine is expanded in a separate turbine called the free (power) turbine. The power turbine drives the propeller through a gear box. Since the power turbine is not connected to the same shaft as the compressor, it can run at any speed depending upon the applied load. The turboshaft engine shown in this figure is quite similar to the turboprop engine, except for the gear box. In a turboshaft engine, the transmission (or gear box) is part of the vehicle, whereas in a turboprop, it is part of the engine. Turboshaft engines are used in helicopters and tanks. In the marine version, the gas turbine is used to drive the propeller(s) of the ship. In the industrial version, the gas turbine is used for power generation. In all the applications shown in Fig. 1.3, the hot gas from the gas turbine exhaust is not required to generate any thrust. Consequently, more enthalpy (and hence more work) can be extracted in the turbine. Hence, in these applications, the turbine has more number of stages to accomplish this.

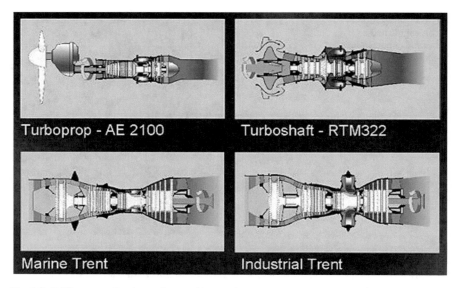

Fig. 1.3 Different applications of gas turbine engine. *Courtesy* Rolls-Royce plc

Chapter 2
Basics of Compressible One-Dimensional Flows

Compressible flows are encountered in many applications in Aerospace and Mechanical engineering. Some examples are flows in nozzles, compressors, turbines and diffusers. In aerospace engineering, in addition to these examples, compressible flows are seen in external aerodynamics, aircraft and rocket engines. In almost all of these applications, air (or some other gas or mixture of gases) is the working fluid. However, steam can be the working substance in turbomachinery applications. Thus, the range of applications in which compressible flow occurs is quite large and hence a clear understanding of the dynamics of compressible flow is essential.

In this chapter, we discuss some fundamental concepts in the study of one-dimensional compressible flows. A flow is said to be one dimensional if the flow properties change only along the flow direction. The fluid can have velocity either along the flow direction or both along and perpendicular to it. Oblique shock waves and the Prandtl–Meyer expansion/compression waves seen in the external intakes of supersonic vehicles are examples of the latter. In addition, the flow can also be quasi one dimensional, as in nozzles and diffusers, wherein the velocity component normal to the flow direction is negligibly small but not zero. The addition of heat to a compressible flow, as for example, in a combustor is also an important interaction. In aerospace propulsion, all these types of flows are encountered and for a detailed discussion of the flow in such applications, the reader is referred to the books given in the list of Suggested Reading.

2.1 Compressibility of Fluids

All fluids are compressible to some extent or other. The compressibility of a fluid is defined as

$$\tau = -\frac{1}{v}\frac{\partial v}{\partial P},\tag{2.1}$$

© The Author(s), under exclusive license to Springer Nature Switzerland AG 2022
V. Babu, *Fundamentals of Propulsion*,
https://doi.org/10.1007/978-3-030-79945-8_2

where v is the specific volume and P is the pressure. The change in pressure corresponding to a given change in specific volume, will, of course, depend upon the compression process. That is, for a given change in specific volume, the change in pressure will be different between an isothermal and an adiabatic compression process.

The definition of compressibility actually comes from thermodynamics. Since the specific volume $v = v(T, P)$, we can write $dv = \left(\frac{\partial v}{\partial P}\right)_T dP + \left(\frac{\partial v}{\partial T}\right)_P dT$. From the first term, we can define the isothermal compressibility as $-\frac{1}{v}\left(\frac{\partial v}{\partial P}\right)_T$, and, from the second term, we can define the coefficient of volume expansion as $\frac{1}{v}\left(\frac{\partial v}{\partial T}\right)_P$. The second term represents change in specific volume (or equivalently density) due to a change in temperature. For example, when a gas is heated at constant pressure, the density decreases and the specific volume increases. This change can be large, as is the case in most combustion equipment, without necessarily having any implications on the compressibility of the fluid.

If the above equation is written in terms of the density ρ, we get

$$\tau = \frac{1}{\rho}\frac{\partial \rho}{\partial P}. \tag{2.2}$$

The isothermal compressibility of water and air under standard atmospheric conditions is $5 \times 10^{-10}\,\text{m}^2/\text{N}$ and $10^{-5}\,\text{m}^2/\text{N}$. Thus, water (in liquid phase) can be treated as an incompressible fluid in all applications. On the contrary, it would seem that air, with compressibility that is five orders of magnitude higher, has to be treated as a compressible fluid in all applications. Fortunately, this is not true when the flow is involved.

2.2 Compressible and Incompressible Flows

It is well known from high school physics that sound (pressure waves) propagates in any medium with speed which depends on the bulk compressibility. The less compressible the medium, the higher the speed of sound. Thus, the speed of sound is a convenient reference speed, when the flow is involved. The speed of sound in the air under normal atmospheric conditions is 330 m/s. The implications of this when there is flow are as follows. Let us say that we are considering the flow of air around an automobile traveling at 120 kph (about 33 m/s). This speed is 1/10th of the speed of sound. In other words, compared with this speed, sound waves travel 10 times faster. Since the speed of sound *appears* to be high compared with the characteristic velocity in the flow field, the medium behaves as though it were incompressible. As the flow velocity becomes comparable to the speed of sound, compressibility effects become more prominent. In reality, the speed of sound itself can vary from one point to another in the flow field and so the velocity at each point has to be compared with the speed of sound at that point. This ratio is called the Mach number, after Ernst

Mach who made pioneering contributions in the study of the propagation of sound waves. Thus, the Mach number at a point in the flow can be written as

$$M = \frac{V}{a},\qquad(2.3)$$

where V is the velocity magnitude at any point and a is the speed of sound at that point.

We can come up with a quantitative criterion to give us an idea about the importance of compressibility effects in the flow by using simple scaling arguments as follows. From Bernoulli's equation for steady flow, it follows that $\Delta P \sim \rho V_{char}^2$, where V_{char} is the characteristic speed. It will be shown later in this chapter that the speed of sound $a = \sqrt{\Delta P/\Delta \rho}$, wherein ΔP and $\Delta \rho$ correspond to an isentropic process. Thus,

$$\frac{\Delta \rho}{\rho} = \frac{1}{\rho}\frac{\Delta \rho}{\Delta P}\Delta P = \frac{V_{char}^2}{a^2} = M^2.\qquad(2.4)$$

On the other hand, upon rewriting Eq. 2.2 for an isentropic process, we get

$$\frac{\Delta \rho}{\rho} = \tau_{isentropic}\Delta P.$$

Comparison of these two equations shows clearly that, in the presence of a flow, density changes are proportional to the square of the Mach number.[1] It is customary to assume that the flow is essentially incompressible if the change in density is less than 10% of the mean value. It thus follows that compressibility effects are significant only when the Mach number exceeds 0.3.

2.3 Perfect Gas Equation of State

In this text, we assume throughout that the working fluid behaves as a perfect gas. The equation of state can be written as

$$Pv = RT,\qquad(2.5)$$

where T is the temperature.[2] R is the particular gas constant and is equal to $\mathcal{R}/\mathcal{M}$, where $\mathcal{R} = 8314\,\text{J/kmol/K}$ is the Universal Gas Constant and $\mathcal{M}$ is the molecular weight in units of kg/kmol. Equation 2.5 can be written in many different forms depending upon the application under consideration. A few of these forms are presented here for the sake of completeness. Since the specific volume $v = 1/\rho$, we can write

---

[1] This is true for steady flows only. For unsteady flows, density changes are proportional to the Mach number.

[2] In later chapters, this will be referred to as the static temperature.

$$P = \rho RT$$

or, alternatively, as

$$PV = mRT,$$

where m is the mass and V is the volume.

2.4 Calorically Perfect Gas

In the study of compressible flows, we need, in addition to the equation of state, an equation relating the internal energy to other measurable properties. The internal energy, strictly speaking, is a function of two thermodynamic properties, namely temperature and pressure. In reality, the dependence on pressure is very weak and hence is usually neglected. Such gases are called *thermally perfect* and for them $e = f(T)$. The exact nature of this function is examined next.

From a molecular perspective, it can be seen intuitively that the internal energy will depend on the number of modes in which energy can be stored (also known as degrees of freedom) by the molecules (or atoms) and the amount of energy that can be stored in each mode. For monatomic gases, the atoms have the freedom to move (and hence store energy in the form of kinetic energy) in any of the three coordinate directions.

For diatomic gases, assuming that the molecules can be modeled as "dumb bells", additional degrees of freedom are possible. These molecules, in addition to translational motion along the three axes, can also rotate about these axes. Hence, energy storage in the form of rotational kinetic energy is also possible. In reality, since the moment of inertia about the "dumb bell" axis is very small, the amount of kinetic energy that can be stored through rotation about this axis is negligible. Thus, rotation adds essentially two degrees of freedom only. In the "dumb bell" model, the bonds connecting the two atoms are idealized as springs. When the temperature increases beyond 600 K or so, these springs begin to vibrate and so energy can now be stored in the form of the vibrational kinetic energy of these springs. When the temperature becomes high (>2000 K), transition to other electronic levels and dissociation take place and at even higher temperatures, the atoms begin to ionize. These effects do not represent degrees of freedom.

Having identified the number of modes of energy storage, we now turn to the amount of energy that can be stored in each mode. The classical equipartition energy principle states that each degree of freedom, when "fully excited", contributes $1/2\,RT$ to the internal energy per unit mass of the gas. The term "fully excited" means that no more energy can be stored in these modes. For example, the translational mode becomes fully excited at temperatures as low as 3 K itself. For diatomic gases, the rotational mode is fully excited beyond 600 K and the vibrational mode beyond 2000 K or so. Strictly speaking, all the modes are quantized and so the energy stored in each mode has to be calculated using quantum mechanics. However, the spacing between the energy levels for the translational and rotational modes are small enough that we can assume the equipartition principle to hold for these modes. We can thus write

$$e = \frac{3}{2}RT$$

for monatomic gases and

$$e = \frac{3}{2}RT + RT + \frac{h\nu/k_B T}{e^{h\nu/k_B T} - 1}RT$$

for diatomic gases. In the above expression, ν is the fundamental vibrational frequency of the molecule. Note that for large values of T, the last term approaches RT. We have not derived this term formally as it would be well outside the scope of this book. Interested readers may see the book by Anderson for full details.

The enthalpy per unit mass can now be calculated by using the fact that

$$h = e + Pv = e + RT.$$

We can calculate C_v and C_p from these equations by using the fact that $C_v = \partial e/\partial T$ and $C_p = \partial h/\partial T$. Thus

$$C_v = \frac{3}{2}R$$

for monatomic gases and

$$C_v = \frac{5}{2}R + \frac{(h\nu/k_B T)^2 \, e^{h\nu/k_B T}}{\left(e^{h\nu/k_B T} - 1\right)^2}R$$

for diatomic gases. The variation of C_v/R is illustrated schematically in Fig. 2.1. It is clear from this figure that $C_v = 5/2R$ in the temperature range $50\,\text{K} \leq T \leq 600\,\text{K}$. In this range $C_p = 7/2R$, and thus the ratio of specific heats $\gamma = 7/5$ for diatomic gases. For monatomic gases, it is easy to show that $\gamma = 5/3$. In this temperature range, where C_v and C_p are constants, the gases are said to be *calorically perfect*. We will assume calorically perfect behavior in all the subsequent chapters. Also, for a calorically perfect gas, since $h = C_p T$ and $e = C_v T$, it follows from the definition of enthalpy that

$$C_p - C_v = R. \tag{2.6}$$

This is called Meyer's relationship. In addition, it is easy to see that

Fig. 2.1 Variation of C_v/R with temperature for diatomic gases

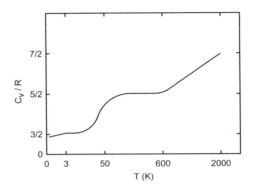

$$C_v = \frac{R}{\gamma - 1}, \qquad C_p = \frac{\gamma R}{\gamma - 1}. \tag{2.7}$$

These relationships will be used extensively throughout in the following chapters.

2.5 Governing Equations

The governing equations for frictionless, adiabatic, steady, one-dimensional flow of a calorically perfect gas can be written in differential form as

$$d(\rho V) = 0, \tag{2.8}$$

$$dP + \rho V dV = 0 \tag{2.9}$$

and

$$dh + d\left(\frac{V^2}{2}\right) = \delta q - \delta w. \tag{2.10}$$

These equations express mass, momentum and energy conservation, respectively. Here, V is the velocity and $\delta q, \delta w$ refer to the heat and work interaction per unit mass. We have also used the customary sign convention from thermodynamics, i.e., that heat added to a system is positive and work done by a system is positive. In addition, changes in flow properties must also obey the second law of thermodynamics. Thus,

$$ds \geq \left(\frac{\delta q}{T}\right)_{rev}, \tag{2.11}$$

where s is the entropy per unit mass and q is the heat interaction, also expressed on a per unit mass basis. The subscript refers to a reversible process. From the first law of thermodynamics, we have

$$de = C_v dT = \delta q_{rev} - P dv. \tag{2.12}$$

Since $\delta q_{rev} = T ds$ from Eq. 2.11 and using the equation of state $Pv = RT$ and its differential form $P dv + v dP = R dT$, we can write

$$ds = C_v \frac{dT}{T} + R \frac{dv}{v} = C_v \frac{dP}{P} + C_p \frac{dv}{v} = C_p \frac{dT}{T} - R \frac{dP}{P}. \tag{2.13}$$

Note that Eq. 2.9 is written in the so-called non-conservative form. By using Eq. 2.8, we can rewrite Eq. 2.9 in conservative form as follows.

$$dP + d\left(\rho V^2\right) = 0. \tag{2.14}$$

Equations 2.8, 2.10, 2.11 and 2.14 can be integrated between any two points in the flow field to give

$$\rho_1 V_1 = \rho_2 V_2, \tag{2.15}$$

$$P_1 + \rho_1 V_1^2 = P_2 + \rho_2 V_2^2, \tag{2.16}$$

$$h_1 + \frac{V_1^2}{2} = h_2 + \frac{V_2^2}{2} - q + w \tag{2.17}$$

and

$$s_2 - s_1 = \int_1^2 \frac{\delta q}{T} + \sigma_{irr}. \tag{2.18}$$

Here, σ_{irr} represents entropy generated due to irreversibilities. It is equal to zero for an isentropic flow and is greater than zero for all other flows. It follows then from Eq. 2.18 that entropy change during an adiabatic process must increase or remain the same. The latter process, which is adiabatic and reversible, is known as an isentropic process. It is important to realize that while all adiabatic and reversible processes are isentropic, the converse need not be true. This can be seen from Eq. 2.18, since with the removal of appropriate amount of heat, the entropy increase due to irreversibilities can be offset entirely (at least in principle), thereby rendering an irreversible process isentropic. Equation 2.18 is not in a convenient form for evaluating entropy change during a process. For this purpose, we can integrate Eq. 2.13 from the initial to the final state during the process. This gives

$$\begin{aligned} s_2 - s_1 &= C_v \ln \frac{T_2}{T_1} + R \ln \frac{v_2}{v_1} \\ &= C_v \ln \frac{P_2}{P_1} + C_p \ln \frac{v_2}{v_1} \\ &= C_p \ln \frac{T_2}{T_1} - R \ln \frac{P_2}{P_1}. \end{aligned} \tag{2.19}$$

Fig. 2.2 Propagation of a
sound wave into a quiescent
fluid

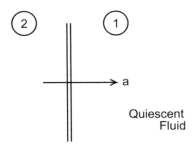

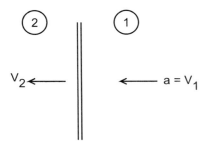

The flow area does not appear in the above equations as they stand. When we discuss one-dimensional flow in ducts and passages, this can be introduced quite easily. Also, it is important to keep in mind when points 1 and 2 are located across a wave (say, a sound wave, shock wave, etc.), the derivatives of the flow properties will be discontinuous.

2.6 Acoustic Wave Propagation Speed

Equations 2.15, 2.16, 2.17 and 2.18 admit different solutions, which we will see in the subsequent chapters. The most basic solution is the expression for the speed of sound, which we will derive in this section.

Consider an acoustic wave propagating into quiescent air as shown in Fig. 2.2. Although the wave front is spherical, at any point on the wave front, the flow is essentially one dimensional as the radius of curvature of the wave front is large when compared with the distance across which the flow properties change. If we switch to a reference frame in which the wave appears stationary, then the flow approaches the wave with a velocity equal to the wave speed in the stationary frame of reference and moves away from the wave with a slightly different velocity. As a result of going through the acoustic wave, the flow properties change by an infinitesimal amount and the process is isentropic. Thus, we can take $V_2 = V_1 + dV$, $P_2 = P_1 + dP$, $\rho_2 = \rho_1 + d\rho$, etc. Substitution of these into Eqs. 2.15 and 2.16 gives

$$\rho_1 V_1 = (\rho_1 + d\rho)(V_1 + dV)$$

and

$$P_1 + \rho_1 V_1^2 = P_1 + dP + (\rho_1 + d\rho)(V_1 + dV)^2.$$

If we neglect the product of differential terms, then we can write

$$\rho_1 dV + V_1 d\rho = 0$$

and

$$dP + 2\rho_1 V_1 dV + V_1^2 d\rho = 0.$$

Upon combining these two equations, we get

$$\frac{dP}{d\rho} = V_1^2.$$

As mentioned earlier, V_1 is equal to the speed of sound a and so

$$a = \sqrt{\frac{dP}{d\rho}}. \qquad (2.20)$$

Since the process is isentropic, $ds = 0$, and so from Eq. 2.13,

$$C_v \frac{dP}{P} + C_p \frac{dv}{v} = 0.$$

Since $\rho = 1/v$, $dv/v = -d\rho/\rho$ and so

$$\frac{dP}{d\rho} = \frac{C_p}{C_v} \frac{P}{\rho} = \gamma R T.$$

Thus,

$$a = \sqrt{\gamma R T} \qquad (2.21)$$

This expression is valid for a non-reacting mixture of ideal gases as well, with the understanding that γ is the ratio of specific heats for the mixture and R is the particular gas constant for the mixture.

2.6.1 Mach Number

The Mach number at a point is defined in Eq. 2.3. Since the Mach number is defined as a ratio, changes in the Mach number are the outcome of either changes in velocity, speed of sound or both. The speed of sound itself varies from point to point and is proportional to the square root of the temperature as seen from Eq. 2.21. Thus, any deductions of the velocity or temperature variation from a given variation of the Mach number cannot be made in a straightforward manner. For example, the velocity at the entry to the combustor in an aircraft gas turbine engine may be as high as 300 m/s, but the Mach number is usually 0.3 or less due to the high static temperature of the fluid.

2.7 Reference States

In the study of compressible flows and indeed in fluid mechanics, it is conventional to define certain reference states. These allow the governing equations to be simplified and written in dimensionless form so that the important parameters can be identified. In the context of compressible flows, the solution procedure can also be made simpler and in addition, the important physics in the flow can be brought out clearly by the use of these reference states. Two such reference states are discussed next.

2.7.1 Sonic State

Since the speed of sound plays a crucial role in compressible flows, it is convenient to use the sonic state as a reference state. The sonic state is the state of the fluid at that point in the flow field where the velocity is equal to the speed of sound. Properties at the sonic state are usually denoted with a *, viz., P^*, T^*, ρ^*, etc. Of course, u^* itself is equal to a and so the Mach number $M = 1$ at the sonic state. The sonic reference state can be thought of as a global reference state since it is attained only at one or a few points in the flow field. For example, in the case of choked isentropic flow through a nozzle, the sonic state is achieved in the throat section. In some other cases, such as flow with heat addition or flow with friction, the sonic state may not even be attained anywhere in the actual flow field, but is still defined in a hypothetical sense and is useful for analysis. The importance of the sonic state lies in the fact that it separates subsonic ($M < 1$) and supersonic ($M > 1$) regions of the flow. Since information travels in a compressible medium through acoustic waves, the sonic state separates regions of flow that are fully accessible (subsonic) and those that are not (supersonic).

Note that the dimensionless velocity V/V^* at a point is **not** equal to the Mach number at that point since V^* is not the speed of sound at that point.[3]

2.7.2 Stagnation State

Let us consider a point in a one-dimensional flow and assume that the state at this point is completely known. This means that the pressure, temperature and velocity at this point are known. We now carry out a thought experiment in which an isentropic, deceleration process takes the fluid from the present state to one with zero velocity. The resulting end state is called the stagnation state corresponding to the known initial state. Thus, the stagnation state at a point in the flow field is defined as the thermodynamic state that would be reached from the given state at that point, at the end of an isentropic, deceleration process to zero velocity. Note that the stagnation state is a local state contrary to the sonic state. Hence, the stagnation state can change from one point to the next in the flow field. Also, it is important to note that the stagnation process alone is isentropic, and the flow need not be isentropic. Properties at the stagnation state are usually indicated with a subscript 0, viz., P_0, T_0, ρ_0, etc. Here P_0 is the stagnation pressure, T_0 is the stagnation temperature and ρ_0 is the stagnation density. Hereafter, P and T will be referred to as the static pressure and static temperature and the corresponding state point will be called the static state.

To derive the relationship between the static and stagnation states, we start by integrating Eq. 2.10 between these two states. This gives

$$\int_1^0 dh + \int_1^0 d\left(\frac{V^2}{2}\right) = 0.$$

If we integrate this equation and rearrange, we get

$$h_0 = h_1 + \frac{V_1^2}{2} \tag{2.22}$$

after noting that the velocity is zero at the stagnation state. For a calorically perfect gas, $dh = C_p dT$ and so

$$T_{0,1} - T_1 = \frac{V_1^2}{2C_p}.$$

After using Eqs. 2.7 and 2.19, we can finally write

_____

[3] Except, of course, at the point where the sonic state occurs.

$$\frac{T_0}{T} = 1 + \frac{\gamma - 1}{2} M^2, \qquad (2.23)$$

where the subscript for the static state has been dropped for convenience. Although the stagnation process is isentropic, this fact is not required for the calculation of stagnation temperature.

Since the stagnation process is isentropic, the static and stagnation states lie on the same isentrope. If we apply Eq. 2.19 between the static and stagnation and use the fact that $s_0 = s_1$, we get

$$\frac{P_{0,1}}{P_1} = \left(\frac{T_{0,1}}{T_1} \right)^{\frac{\gamma}{\gamma - 1}}.$$

If we substitute from Eq. 2.23, we get

$$\frac{P_0}{P} = \left(1 + \frac{\gamma - 1}{2} M^2 \right)^{\frac{\gamma}{\gamma - 1}}, \qquad (2.24)$$

where the subscript denoting the static state has been dropped. This equation can be derived in an alternative way in a manner similar to the one used for the derivation of the stagnation temperature. This is somewhat longer but gives some interesting insights into the stagnation process. We start by rewriting Eq. 2.9 in the following form.

$$\frac{dP}{\rho} + d\left(\frac{V^2}{2} \right) = 0.$$

By substituting Eq. 2.10, this can be simplified to read

$$\frac{dP}{\rho} - dh = 0.$$

Integrating this between the static and stagnation states leads to

$$\int_1^0 \frac{dP}{\rho} - \int_1^0 dh = 0.$$

Since the second term is a perfect differential, it can be integrated easily. The first term is not a perfect differential and so the integral depends on the path used for the integration—in other words, the path connecting states 1 and 0. Since this process is isentropic, from Eq. 2.13, we can show that

$$C_v \frac{dP}{P} + C_p \frac{dv}{v} = 0 \Rightarrow Pv^\gamma = \text{constant} = P_1 v_1^\gamma.$$

Thus, the above equation reduces to

$$\int_1^0 \frac{P_1^{1/\gamma}}{\rho_1} \frac{dP}{P^{1/\gamma}} = C_p(T_{0,1} - T_1),$$

where we have invoked the calorically perfect gas assumption. With a little bit of algebra, this can be easily shown to lead to Eq. 2.24.

The stagnation density can be evaluated by using the equation of state $P_0 = \rho_0 R T_0$. Thus,

$$\frac{\rho_0}{\rho} = \left(1 + \frac{\gamma - 1}{2} M^2\right)^{\frac{1}{\gamma-1}}. \tag{2.25}$$

This derivation brings out the fact that unlike the stagnation temperature, the nature of the stagnation process has to be known in order to evaluate the stagnation pressure. This, in itself, arises from the fact that Eq. 2.9 is not a perfect differential.

Another important fact about stagnation quantities is that they depend on the frame of reference unlike static quantities which are frame independent. This is best illustrated through a numerical example.

Example 2.1 Consider the propagation of sound wave into quiescent air at 300 K and 100 kPa. With reference to Fig. 2.1, determine $T_{0,1}$ and $P_{0,1}$ in the stationary and moving frames of reference.

Solution: In the stationary frame of reference, $V_1 = 0$ and so $T_{0,1} = T_1 = 300$ K and $P_{0,1} = P_1 = 100$ kPa.

In the moving frame of reference, $V_1 = a_1$ and so $M_1 = 1$. Substituting this into Eqs. 2.23 and 2.24, we get $T_{0,1} = 360$ K and $P_{0,1} = 189$ kPa.

The difference between the values evaluated in different frames becomes more pronounced at higher Mach numbers.

As already mentioned, stagnation temperature and pressure are local quantities and so they can change from one point to another in the flow field. Changes in stagnation temperature can be achieved by the addition or removal of heat or work. Heat addition increases the stagnation temperature, while removal of heat results in a decrease in stagnation temperature. Changes in stagnation pressure are brought about by work interaction or irreversibilities. Across a compressor where work is done on the flow, stagnation pressure increases while across a turbine where work is extracted from the fluid, stagnation pressure decreases. It is for this reason that any loss of stagnation pressure in the flow is undesirable as it is tantamount to loss of work. To see the effect of irreversibilities, we start with the last equality in Eq. 2.19 and substitute for T_2/T_1 and P_2/P_1 as follows:

$$\frac{T_2}{T_1} = \frac{T_2}{T_{0,2}} \frac{T_{0,1}}{T_1} \frac{T_{0,2}}{T_{0,1}}$$

and

$$\frac{P_2}{P_1} = \frac{P_2}{P_{0,2}} \frac{P_{0,1}}{P_1} \frac{P_{0,2}}{P_{0,1}}.$$

From Eq. 2.19,

$$s_2 - s_1 = R \ln \left[\left(\frac{T_2}{T_1}\right)^{\frac{\gamma}{\gamma-1}} \bigg/ \left(\frac{P_2}{P_1}\right) \right].$$

If we use Eqs. 2.23 and 2.24, we get

$$s_2 - s_1 = C_p \ln \frac{T_{0,2}}{T_{0,1}} - R \ln \frac{P_{0,2}}{P_{0,1}}. \tag{2.26}$$

This equation shows that irreversibilities in an adiabatic flow lead to a loss of stagnation pressure, since, for such a flow, $s_2 > s_1$ and $T_{0,2} = T_{0,1}$ and so $P_{0,2} < P_{0,1}$. This equation also shows that heat addition in a compressible flow is always accompanied by a loss of stagnation pressure. Since $T_{0,2} > T_{0,1}$, in this case, and $s_2 > s_1$, $P_{0,2}$ has to be less than $P_{0,1}$. These facts are important in the design of combustors and will be discussed later. This equation also shows that the increase or decrease of stagnation pressure brought about through work interaction leads to a corresponding change in the stagnation temperature.

2.8 T-s Diagrams in Compressible Flows

T-s and P-v diagrams are familiar to most of the readers from their basic thermodynamics course. These diagrams are extremely useful in illustrating states and processes graphically. Both of these diagrams display the same information since the thermodynamic state is fully fixed by the specification of two properties, either P, v or T, s. Nevertheless, they are both useful as some processes can be depicted better in one than the other. In the context of propulsion, T-s diagrams are more useful and widely used.

Let us review some basic concepts from thermodynamics in relation to T-s diagram. Figure 2.3 shows thermodynamic states (filled circles) and contours of T, s (isotherms and isentropes). From the first equality in Eq. 2.13, we can write

$$dv = \frac{v}{R} ds - C_v \frac{v}{RT} dT. \tag{2.27}$$

From this equation, it is easy to see that as we move along a $s = $ constant line in the direction of increasing temperature, v decreases, since $dv = -(C_v/P)dT$ along such a line. Also, the change in v for a given change in T is higher at lower values of pressure than at higher values of pressure. This fact is of tremendous importance in compressible flows as we will see later.

Fig. 2.3 Constant pressure
and constant volume lines on
a T-s diagram

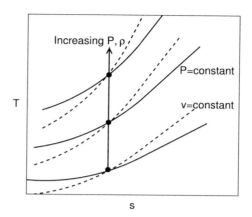

Since $dv = 0$ along a $v = $ constant contour, from the above equation,

$$\frac{dT}{ds} = \frac{T}{C_v}.$$

(2.28)

This equation shows that the slope of the contours of v on a T-s diagram is always positive and is not a constant. Hence, the contours are not straight lines. Furthermore, the slope increases with increasing temperature and so the contours are horizontal at low temperatures and become steeper at higher temperatures.

Similarly, from the third equality in Eq. 2.13, it can be shown that

$$\frac{dT}{ds} = \frac{T}{C_p}$$

(2.29)

for any isobar. The same observation made above regarding the slope of the contours of v on a T-s diagram is applicable to isobars as well. In addition, since $C_p > C_v$, at any state point, isochors are steeper than isobars on a T-s diagram and pressure increases along a $s = $ constant line in the direction of increasing temperature. These observations regarding isochors and isobars are shown in Fig. 2.3.

Let us now look at using T-s diagram for graphically illustrating states in 1D compressible flows. In this case, in addition to T, s, velocity information also has to be displayed. Equation 2.10 tells us how this can be done. For a calorically perfect gas, this can be written as

$$d\left(T + \frac{V^2}{2C_p}\right) = 0.$$

Hence, at each state point, the static temperature is depicted as usual, and the quantity $V^2/2C_p$ is added to the ordinate (in the case of a T-s diagram). Note that this quantity has units of temperature and the sum $T + V^2/2C_p$ is equal to the stagnation

Fig. 2.4 Illustration of states
for a 1D compressible flow
on T-s and P-v diagram

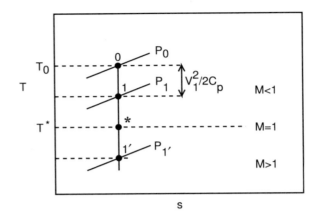

temperature T_0 corresponding to this state. This is shown in Fig. 2.4 for the subsonic state point marked 1. Also shown in this figure is the sonic state *corresponding to this state*. Once T_0 is known, T^* can be evaluated from Eq. 2.23 by setting $M = 1$. Thus,

$$\frac{T_0}{T^*} = \frac{\gamma + 1}{2}.$$

Depicting the sonic state is useful since it tells at a glance whether the flow is subsonic or supersonic. All subsonic states will lie above the sonic state and all supersonic states will lie below. State point $1'$ shown in Fig. 2.3 is a supersonic state. This figure also shows that the stagnation process (1-0 or $1'$-0) is an isentropic process.

Exercises

2.1 Air enters the diffuser of an aircraft jet engine at a static pressure of 20 kPa, static temperature 217 K and a Mach number of 0.9. The air leaves the diffuser with a velocity of 85 m/s. Assuming isentropic operation, determine the exit static temperature and pressure.
[249 K, 32 kPa]

2.2 Air is compressed adiabatically in a compressor from a static pressure of 100–2000 kPa. If the static temperature of the air at the inlet and exit of the compressor is 300 and 800 K, determine the power required per unit mass flow rate of air. Also, determine whether the compression process is isentropic or not.
[503 kW, Not isentropic]

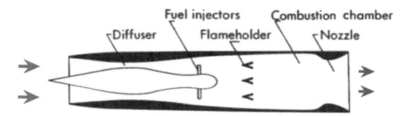

Fig. 2.5 Schematic of a ramjet engine

2.3 Air enters a turbine at a static pressure of 2 MPa, 1400 K. It expands isentropically in the turbine to a pressure of 500 kPa. Determine the work developed by the turbine per unit mass flow rate of air and the static temperature at the exit.
[460 kW, 942 K]

2.4 Air at 100 kPa, 295 K and moving at 710 m/s is decelerated isentropically to 250 m/s. Determine the final static temperature and static pressure.
[515 K, 702 kPa]

2.5 Air enters a combustion chamber at 150 kPa, 300 K and 75 m/s. Heat addition in the combustion chamber amounts to 900 kJ/kg. Air leaves the combustion chamber at 110 kPa and 1128 K. Determine the stagnation temperature, stagnation pressure and velocity at the exit and the entropy change across the combustion chamber.
[1198 K, 115 kPa, 168 m/s, 1420 J/kg K]

2.6 Air at 900 K and negligible velocity enters the nozzle of an aircraft jet engine. If the flow is sonic at the nozzle exit, determine the exit static temperature and velocity. Assume adiabatic operation.
[750 K, 549 m/s]

2.7 Air expands isentropically in a rocket nozzle from $P_0 = 3.5$ MPa, $T_0 = 2700$ K to ambient pressure of 100 kPa. Determine the exit velocity, Mach number and static temperature.
[1860 m/s, 2.97, 978 K]

2.8 The ramjet engine shown in Fig. 2.5 does not have any moving parts. It operates at high supersonic Mach numbers (<4). The entering is decelerated in the diffuser to subsonic speed. Heat is added to the combustion chamber and the hot gases expand in the nozzle generating thrust. In an "ideal" ramjet engine, the air is the working fluid throughout and the compression and expansion processes are isentropic. In addition, there is no loss of stagnation pressure due to the heat addition. The air is expanded in the nozzle to the ambient pressure. Show that the Mach number of the air as it leaves the nozzle is the same as the Mach number of the air when it enters the diffuser. Sketch the process undergone by the air on a T-s diagram.

Chapter 3
Basics of Turbomachinery

Since the compressor, fan and turbine are an integral part of any gas turbine-based engine, a clear understanding of the fundamental fluid- and thermo-dynamic processes occurring in such turbomachines is essential. With this in mind, some basic concepts relating to turbomachines are discussed in this chapter. For a detailed discussion on these and other advanced topics, the reader is referred to the classic treatise on turbomachinery by Shepherd.

3.1 Euler's Turbomachinery Equation

Consider the rotor of a turbomachine in which the fluid enters at a radius r_1 with absolute velocity V_1 and leaves at radius r_2 with absolute velocity V_2. Let the angular velocity of the rotor be ω radians/sec. The absolute velocity can be resolved into three mutually perpendicular components denoted by V_x, V_r and V_θ. Here, the subscript x denotes the axial direction (pointing along the axis of rotation, from inlet to exit), r denotes the radial direction (pointing outward from the axis) and θ denotes the tangential direction (pointing along the direction of rotation). The tangential component V_θ is also called the whirl velocity. The sign convention for the velocity components is the same as those of their respective axes. The torque (which acts along the θ direction) on the rotor shaft is given as

$$\text{Torque} = \dot{m}\left(V_{\theta,1}\, r_1 - V_{\theta,2}\, r_2\right),$$

where $\dot{m}$ is the mass flow rate through the rotor. If we multiply both sides of this equation by ω and use the fact that torque multiplied by angular velocity equals power, we get

$$\dot{W} = \dot{m}\left(V_{\theta,1}\, \omega r_1 - V_{\theta,2}\, \omega r_2\right).$$

© The Author(s), under exclusive license to Springer Nature Switzerland AG 2022
V. Babu, *Fundamentals of Propulsion*,
https://doi.org/10.1007/978-3-030-79945-8_3

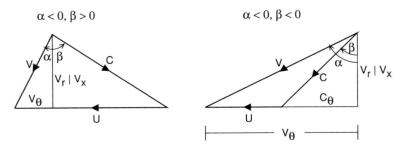

Fig. 3.1 General velocity triangle for a rotor

Here, $\dot{W}$ represents the power or rate at which work is done. Note that the sign convention for $\dot{W}$ is the same as that of thermodynamic work, i.e., work done by the fluid is positive. Since the product $r\omega$ is equal to the tangential velocity of the *rotor* (denoted by U) at a radius r, the above expression can be written as

$$\dot{W} = \dot{m}\left(V_{\theta,1}U_1 - V_{\theta,2}U_2\right).\tag{3.1}$$

This equation is called the Euler turbine equation and is of fundamental importance in the theory of turbomachinery. The difference in the magnitude of the product $V_\theta U$ at the inlet and outlet determines whether the machine is work producing (turbine) or work absorbing (compressor, pump, fan, blower, etc.).

Alternative forms of this equation are also very useful. Before we go on to derive one such alternative form, we need to have an understanding of the relationship between different velocities that occur in a turbomachine. Figure 3.1 illustrates geometrically the different velocities for the case of a general turbomachine. In the case of an axial flow machine, only the axial and tangential velocities are of interest and hence it is sufficient to consider these components on an $r =$ constant plane. In the case of a radial flow machine, the radial and tangential components of velocities on an $x =$ constant plane alone are of interest. Two velocity triangles are shown in Fig. 3.1, one corresponding to $V_\theta < U$ and the other to $V_\theta > U$. In Fig. 3.1, C is the velocity of the fluid relative to the rotor, $\mathbf{C} = \mathbf{V} - \mathbf{U}$. In other words, an observer sitting on the rotor at this radius would see the fluid approaching the rotor with velocity C. The relative velocity vectors at the inlet and outlet are such that the fluid would appear to glide on and off the rotor blade (i.e., *tangential to the blade*) to this observer.[1] Since the tangential velocity of the rotor changes with radius (from root to tip), the relative velocity C will also be different at different radii. This suggests that the rotor blade angles will also change with radius to accommodate the changing relative velocity. The flow angle and blade angle measured with respect to a chosen reference direction are indicated in Fig. 3.1 as α and β, respectively. In the case of an axial flow rotor, the

---

[1] We have assumed here, without any loss of generality, that the angle of attack at the inlet and the angle of deviation at the exit for the relative velocity vector are both equal to zero.

reference direction is along the axis. The sign convention is that an angle measured anti-clockwise from the reference direction is positive.

From the geometric construction of the velocity triangle in Fig. 3.1, it can be seen that $C_r = V_r$, $V^2 = V_r^2 + V_\theta^2$ and $C^2 = V_r^2 + C_\theta^2$ for a radial flow machine. Also, in the general case, $V_\theta = U \pm C_\theta$. Therefore, $V_\theta U = U^2 \pm C_\theta U$. From the first relation, we get

$$V^2 = C^2 - C_\theta^2 + (U \pm C_\theta)^2 = C^2 + U^2 \pm 2U C_\theta.$$

Thus,

$$\pm 2U C_\theta = V^2 - C^2 - U^2$$

and

$$V_\theta U = U^2 + \frac{1}{2}\left(V^2 - C^2 - U^2\right) = \frac{1}{2}\left(V^2 + U^2 - C^2\right).$$

Upon substituting this into Eq. 3.1, we get

$$\dot{W} = \dot{m}\left(\frac{V_1^2 - V_2^2}{2} + \frac{U_1^2 - U_2^2}{2} - \frac{C_1^2 - C_2^2}{2}\right). \tag{3.2}$$

Although we have derived this equation for a radial flow rotor, it is applicable equally well for axial and mixed flow rotors also. It is left as an exercise for the reader to demonstrate this fact.

Since our interest is primarily on aerospace propulsion, turbomachines used for this application are almost exclusively axial flow machines (the rationale for this is discussed later). Accordingly, we will confine our discussion to axial flow machines from here onward. For an axial machine, $U_1 = U_2 = U$ and so Eq. 3.1 becomes

$$\dot{W} = \dot{m}U\left(V_{\theta 1} - V_{\theta 2}\right). \tag{3.3}$$

3.2 Relationship Between the Fluid Mechanics and Thermodynamics of Turbomachinery Process

So far we have looked at the fluid mechanical aspects of the flow through a turbomachinery rotor and developed an expression for the power developed or absorbed by the rotor in terms of velocities alone. In addition, we also need to relate the power to changes in thermodynamic properties of the fluid, such as temperature and pressure. To this end, we substitute for w in Eq. 2.17 from Eq. 3.2, after noting that $w = \dot{W}/\dot{m}$. This leads to

$$h_1 - h_2 + \frac{V_1^2 - V_2^2}{2} = \frac{V_1^2 - V_2^2}{2} + \frac{U_1^2 - U_2^2}{2} - \frac{C_1^2 - C_2^2}{2},$$

where we have assumed the energy transfer process in the rotor to be adiabatic. This equation can be used to determine the change in the stagnation or static enthalpy across the rotor. Thus,

$$h_{01} - h_{02} = \frac{V_1^2 - V_2^2}{2} + \frac{U_1^2 - U_2^2}{2} - \frac{C_1^2 - C_2^2}{2} \tag{3.4}$$

or

$$h_1 - h_2 = \frac{U_1^2 - U_2^2}{2} - \frac{C_1^2 - C_2^2}{2}. \tag{3.5}$$

For an axial machine, the above two equations can be simplified to read

$$h_{0,1} - h_{0,2} = \frac{V_1^2 - V_2^2}{2} - \frac{C_1^2 - C_2^2}{2} \tag{3.6}$$

and

$$h_1 - h_2 = -\frac{C_1^2 - C_2^2}{2}. \tag{3.7}$$

Equation 3.7 can be rearranged to give

$$h_1 - h_2 + \frac{C_1^2 - C_2^2}{2} = 0$$

or

$$h_1 + \frac{C_1^2}{2} = h_2 + \frac{C_2^2}{2} = h_{0,\text{relative}}. \tag{3.8}$$

These can be recognized as the energy equation Eq. 2.17 for an observer moving with the rotor. It is important to note that there is no work interaction in this frame of reference and hence there is no change in the relative stagnation enthalpy. Equations 3.6, 3.7 and 3.8 are illustrated graphically on T-s diagram in Fig. 3.2. Here, we have assumed in addition that the process is reversible. However, irreversibilities can also be handled quite easily within the framework that has been developed here.

3.3 Blade Design Parameters

3.3.1 Flow and Loading Coefficients

The magnitude of $V_{\theta 1} - V_{\theta 2}$ is called the *turning* or *deflection* of the flow as it passes through the rotor. This gives an indication of the amount of work transfer that occurs in the rotor. This is easily seen by writing Eq. 3.3 as $\dot{W} = F_\theta U$, where F_θ is the tangential force that acts on the rotor. For a given mass flow rate, when the flow turning or deflection is high, the work transfer is also high and such rotors are said

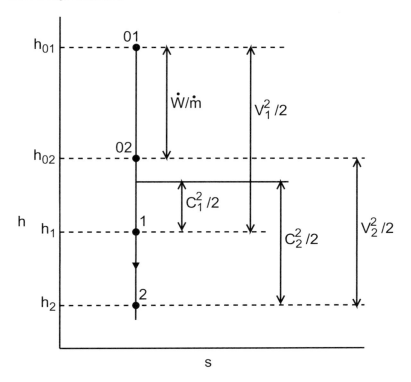

Fig. 3.2 Illustration of isentropic expansion in an axial turbine rotor

to be *heavily loaded*. This notion naturally leads to the definition of a *blade loading coefficient*,

$$\psi = \frac{\dot{W}/\dot{m}}{U^2} = \frac{F_\theta}{\dot{m}U} = \frac{V_{\theta,1} - V_{\theta,2}}{U}. \tag{3.9}$$

Thus, ψ is a measure of the tangential force (which arises due to a difference in pressure force on either side of the blade) that acts on the blade. Let us now see how the blade coefficient ψ affects the geometry of the blade (and hence the flow passage) in the rotor.

Figure 3.3 illustrates a few blade passages in an axial turbine and axial compressor rotor. Several important things can be discerned from this figure, namely:

- The cross-sectional area of the blade passage (indicated by dashed line) appears to decrease in the case of a turbine and increase in the case of a compressor.
- Consequently, as illustrated in this figure, the magnitude of the relative velocity increases from the inlet to exit in the former and decreases in the latter.
- It follows that the flow undergoes expansion (due to flow acceleration) in the former and compression (due to flow deceleration) in the latter.

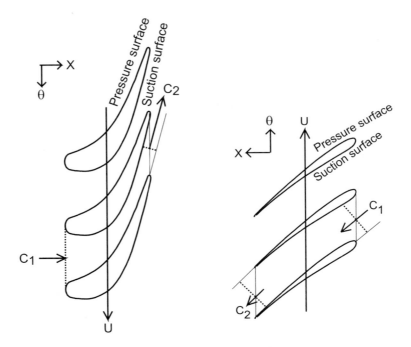

Fig. 3.3 Blade passage of an axial turbine rotor (left) and axial compressor rotor (right)

- Hence, the blade passage of a turbine resembles a nozzle and that of a compressor, a diffuser.
- The flow turning and hence the blade loading coefficient is high in the case of the turbine. The turbine blades are consequently thicker. In the case of the compressor, the flow turning and so the blade loading coefficient are quite small and the blades are also slender.

Velocity triangles at the inlet and exit of the rotor blade passage for the turbine and compressor blades illustrated in Fig. 3.3 are given in Figs. 3.4 and 3.5. These offer further insights into the nature of the fluid mechanics process in the blade passage. In these figures, the combined velocity diagram is also shown by using the blade velocity vector as a common base. It is also possible to draw the combined diagram to have a common apex. It should be noted that C_1 and C_2 are tangential to the blade profiles at inlet and exit, while V_1 and V_2 are not. The change in velocity of the fluid is also illustrated in these figures as the vector $\Delta\mathbf{V} = \mathbf{V}_2 - \mathbf{V}_1$. Since $V_{x,1} = V_{x,2}$ in these figures (which need not always be true), $\Delta\mathbf{V}$ is parallel to $\mathbf{U}$. The rate of change of momentum of the fluid, $\dot{m}\Delta\mathbf{V}$, which is equal to the force exerted on the fluid by the rotor, acts along the same direction. It should also be noted from Figs. 3.4 and 3.5 that $\Delta\mathbf{V}_\theta$ is in the same direction as U for a turbine and exactly opposite for

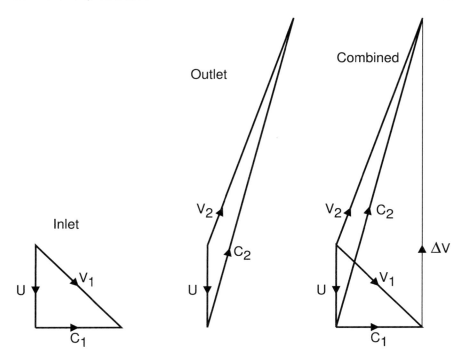

Fig. 3.4 Velocity triangles at the inlet and exit of an axial turbine rotor

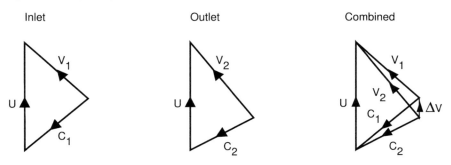

Fig. 3.5 Velocity triangles at the inlet and exit of an axial compressor rotor

a compressor.[2] The tangential component of this force F_θ generates torque and the axial component F_x generates an axial thrust force on the rotor shaft. Thus, it is clear from these figures that the blade loading coefficient is high for the turbine blade and

---

[2] From Newton's third law, the force exerted by the fluid on the rotor is equal and opposite in direction. Thus, for a turbine since work is done by the fluid, the rotor turns in a direction opposite to ΔV_θ. Whereas in a compressor since work is done on the fluid, the direction of rotation is the same as ΔV_θ. The direction of the force exerted on the blade by the fluid can also be determined from the indicated pressure and suction surfaces.

low for the compressor blade. It follows then that the pressure drop in the turbine blade passage is high, while the pressure rise in the blade passage of the compressor is modest. Of course, a blade passage can be designed to have any value for blade loading coefficient (and thus pressure change) but practical considerations impose restrictions, particularly for a compressor. This is because the pressure gradient in the flow direction is favorable in the case of a turbine and adverse in the case of a compressor. Since the boundary layers on the blade surface are prone to separation in an adverse pressure gradient, the pressure rise (and so the blade loading coefficient) that can be realized in a compressor blade passage is small. This inherent limitation in an axial compressor is explored in detail in Chap. 5.

The flow coefficient ϕ for an axial flow machine is defined as the ratio V_x/U. Judging from the velocity triangles shown in Figs. 3.4 and 3.5, it is clear that the flow coefficient determines the distance of the apex of the velocity triangle in proportion to its base. The importance of the flow coefficient is its relationship to the mass flow rate, i.e., $\dot{m} = \rho_1 A_{x,1} V_{x,1} = \rho_2 A_{x,2} V_{x,2}$ (here, we have neglected any variation in flow properties from the root to the tip of the blade). A_x is the cross-sectional area of the rotor on an $x =$ constant plane. In fact, if the velocity triangles in Figs. 3.4 and 3.5 are plotted after dividing all the velocities by U, the resulting velocity triangle is called a normalized velocity triangle. The base of this triangle will have unit length, the altitude will be equal to ϕ and ΔV_θ will be equal to ψ.

3.3.2 Degree of Reaction

Let us now take a closer look at Eqs. 3.6 and 3.7 and get some insights into the energy transfer process in the rotor. As there is definitely work interaction in a rotor, there has to be a change in the stagnation enthalpy of the fluid, as we learnt in the previous chapter. Since the stagnation enthalpy is the sum of the static enthalpy and the velocity squared term, i.e., $h_0 = h + V^2/2$, the change in h_0 can be due to a change in the static enthalpy, h, or a change in the absolute velocity of the fluid, V, or both. The contribution of the change in static enthalpy to the overall change in stagnation enthalpy is called the degree of reaction $\mathcal{R}$,

$$\mathcal{R} = \frac{h_1 - h_2}{h_{0,1} - h_{0,2}}. \tag{3.10}$$

Stated another way, a given amount of work transfer (or equivalently change in h_0) can be achieved through a change in static enthalpy or change in absolute velocity or both. From Eq. 3.7, it can be seen that any change in static enthalpy of the fluid is necessarily accompanied by a change in the relative velocity. Thus, a given amount of work transfer can be accomplished through a change in relative velocity or absolute velocity or both. We have already seen that any change in the magnitude of the relative velocity in the rotor passage would mean that there is a change in static pressure across the rotor and also that the cross-sectional area of the

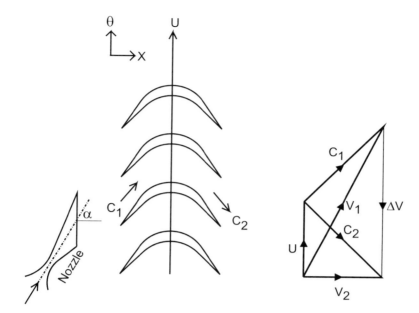

Fig. 3.6 Pure impulse turbine and the corresponding combined velocity triangle

rotor passage changes from inlet to exit. It thus becomes clear that the turbine and compressor blades illustrated in Fig. 3.3 have a non-zero degree of reaction. Such blades are called reaction blades.

In contrast, the rotor blades can be designed to result in a blade passage of constant cross-sectional area, as shown in Fig. 3.6. There is no change in magnitude of the relative velocity of the fluid across the rotor in this case, and so there is no change in static enthalpy and static pressure across the rotor. Such a machine is called an impulse machine. A well-known example of such a machine is the water wheel or the Pelton wheel. Since there is no pressure change across the rotor, the rotor need not be enclosed in a casing at all—it can be open to the atmosphere. The impulse machine shown in Fig. 3.6 and indeed any impulse machine generate power due to a change in the **direction** of the relative velocity and not magnitude. However, there is a change in the direction and magnitude (decrease) of the absolute velocity. Although there is no change in pressure (or enthalpy) of the fluid in the rotor itself, the high initial velocity of the fluid is achieved in a nozzle outside the rotor as a result of a large pressure drop.

From Eq. 3.6, it is easy to see that, for a given amount of work transfer, the change in absolute velocity is higher for an impulse machine compared to a reaction machine. This usually results in higher blade speeds for the impulse machine. Reaction machines, on the contrary, achieve work transfer through a combination of change in relative and absolute velocity and so they tend to be more compact and also run at lower blade speeds.

Fig. 3.7 Illustration for worked Example 3.1

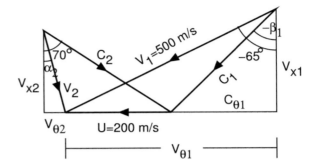

Example 3.1 Steam enters the rotor of an impulse turbine at 500 m/s. The blade speed is 200 m/s. The nozzle angle is $-65°$ and the blade angle at exit is 70°. If the mass flow rate is 1.8 kg/s and the relative velocity remains constant, determine the blade angle at the inlet, the absolute velocity and flow angle at the exit, the power produced and the axial thrust.

Solution: The given information is displayed graphically in Fig. 3.7.

Starting with the inlet, we can get

$V_{\theta,1} = V_1 \sin 65 = 453.15$ m/s, $C_{\theta,1} = V_{\theta,1} - U = 253.15$ m/s and $V_{x1} = V_1$
$\cos 65 = 211.31$ m/s $= C_{x1}$.

Since $\tan \beta_1 = C_{\theta,1}/C_{x,1}$, $\beta_1 = 50.15°$ and $C_1 = 329.75$ m/s $= C_2$.

On the outlet side

$C_{\theta,2} = C_2 \sin \beta_2 = 309.87$ m/s and $C_{x,2} = C_2 \cos \beta_2 = 112.78$ m/s $= V_{x2}$.
$V_{\theta,2} = C_{\theta,2} - U = 109.87$ m/s. Therefore,

$V_2 = \sqrt{V_{\theta,2}^2 + V_{x,2}^2} = 157.45$ m/s and $\alpha_2 = \tan^{-1}(V_{\theta,2}/V_{x,2}) = 44.25°$.

$\dot{W} = \dot{m}U(V_{\theta,1} - V_{\theta,2}) = (1.8)(200)(453.15 + 109.87) = 202.7$ kW.

Note that $V_{\theta 1}$ and $V_{\theta 2}$ are opposite in direction.

Axial thrust $= \dot{m}(V_{x,1} - V_{x,2}) = (1.8)(211.31 - 112.78) = 177.354$ N.

3.4 Multi-stage Turbomachines

We have discussed the fluid- and thermo-dynamic process occurring across a single rotor blade in a turbomachine. In real-life applications, however, it is impractical to have the entire work transfer take place in a single rotor blade. The rotational speed and the blade loading coefficient and the consequent mechanical stresses on the rotor will simply be too high. It is practical to have the work transfer take place in increments across several rows of rotor blades. One difficulty with this arrangement is the fact that the fluid angles at the entry and exit of a rotor blade are not the same (see Figs. 3.4, 3.5 and 3.6). This can be overcome by having a row of stationary blades (called stator blades or guide vanes) in between the rotor blade rows to guide the

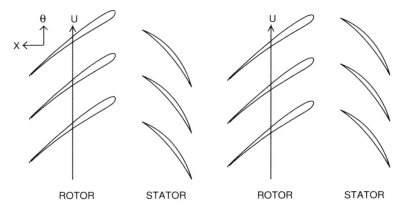

Fig. 3.8 Illustration of two stages of a multi-stage axial compressor

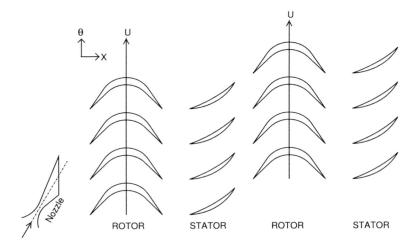

Fig. 3.9 Illustration of two stages of a multi-stage axial turbine

fluid properly from the exit of one rotor blade row to the inlet of the next one. Each stator–rotor pair is called a stage and this design strategy is known as multi-staging. Figures 3.8 and 3.9 illustrate two stages of a multi-stage compressor and turbine, respectively, each starting from the corresponding rotor blade geometry shown in Figs. 3.3 and 3.6.

The moving blades are attached to the rotor at the hub, which in turn is connected to the shaft, while the stator blades are fixed to the casing. As discussed earlier, since the blade loading coefficient can be high for a turbine stage, for a given amount of work transfer, the number of turbine stages is less than the number of compressor stages. For example, in aircraft gas turbine engines, the number of turbine stages

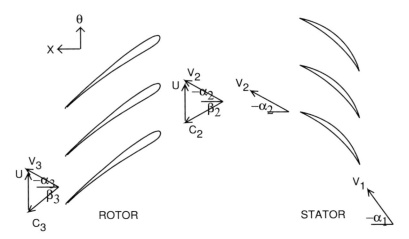

Fig. 3.10 Illustration of a single stage of a multi-stage axial compressor

will be less than 10 and the compressor which is driven by the turbine usually has as many as 20 stages.

The shape of the stator blades are designed based on the absolute velocity vector at the exit and inlet of the rotor blade. The blades themselves can be arranged to achieve just a change in the direction of the fluid velocity, in which case the blade passage area is constant, or a change in the static enthalpy (or pressure) through a varying area passage. In the latter case, the stator blade passage will act as a nozzle in a turbine and as a diffuser in a compressor. There will not be a change in the stagnation enthalpy of the fluid in the stator passage as there is no work interaction. Consequently, for multi-stage machines, the degree of reaction is defined for a stage rather than the rotor alone. With reference to the stage nomenclature used in Fig. 3.10,

$$R = \frac{h_2 - h_3}{h_{0,1} - h_{0,3}} = \frac{h_2 - h_3}{h_{0,2} - h_{0,3}}. \tag{3.11}$$

The second equality follows from the fact that the stagnation enthalpy remains constant across the stator. There are actually several definitions in use for the degree of reaction. The definitions given in Eqs. 3.10 and 3.11 are based on energy considerations. Alternately, the degree of reaction for a stage can also be defined as

$$R = \frac{h_2 - h_3}{h_1 - h_3} \tag{3.12}$$

or

$$R = \frac{P_2 - P_3}{P_1 - P_3}. \tag{3.13}$$

For an incompressible fluid such as water for which the density changes very little, it follows from $h = u + P/\rho$ that $\Delta h = \Delta P/\rho$ (assuming that there is no change in the temperature of the fluid as a result of passage through the stage). The above two definitions for the degree of reaction are thus equivalent in this case. However, for a compressible fluid, it should be noted that a zero reaction rotor and an impulse rotor are not the same. In the former, there is no change in static enthalpy in the rotor and in the latter, there is no change in the static pressure in the rotor. It is generally accepted that an impulse machine is one in which there is no pressure change in the rotor. If the stage is a "normal" stage for which $V_3 = V_1$ and $\alpha_3 = \alpha_1$, then Eqs. 3.11 and 3.12 become identical.

Contrary to the degree of reaction for the rotor, which can take on values between 0 and 1, the degree of reaction for a stage, in addition, can be negative or greater than 1. The latter two situations, although not of practical interest, offer some insight into the nature of the energy transfer process in a stage. They are easy to visualize with the definition of $\mathcal{R}$ given in Eq. 3.13. In general, there is continuous diffusing action (pressure rise) across the stator and rotor in a compressor and expansion (pressure drop) in the case of a turbine. Let us assume, for example, that the stator acts as a nozzle in a compressor, expanding the flow instead of diffusing it. Then, for the stage to have an overall pressure rise, the pressure rise in the rotor has to be higher than the stage pressure rise. This will result in a value for $\mathcal{R}$ greater than 1. On the other hand, in a turbine rotor if $C_3 > C_2$, then $\mathcal{R}$ will be negative, since $h_2 - h_3$ is negative (from Eq. 3.7). In practical applications, the impulse stage and 50% reaction stage are widely used.

Another issue that must be dealt with is the change in density of the fluid as it goes through the compressor or turbine stages. Since the mass flow rate, $\dot{m} = \rho A_x V_x$ is constant from one stage to another, this necessitates a change in cross-sectional area or axial velocity or both. Constant axial velocity is generally preferred since this design allows the flow to move through the stages uniformly. The blade shapes illustrated in Figs. 3.3 and 3.6 are designed to have constant axial velocity as can be ascertained from the corresponding velocity triangles. It follows then that in a compressor the cross-sectional area must decrease along the flow direction (since the density increases as the flow is compressed) and must increase in a turbine (since the density decreases as the flow expands). The change in the cross-sectional area can be accomplished either through constant casing diameter and varying rotor diameter or varying casing diameter and constant rotor diameter. In aircraft engine application, the overriding concern is the frontal area of the engine and the drag associated with it. Hence, constant casing diameter designs are widely used. The interested reader can see the book by Saravanamuttoo, Rogers and Cohen for further details.

Example 3.2 In a reaction turbine, the rotor and stator blades are the same shape but are reversed in direction. The inlet and exit angles are 55° and 70°, respectively (see figure). Show that the degree of reaction in this case is 50%. If the blade speed is 40 m/s and the mass flow rate through the turbine is 1 kg/s, determine the power developed, flow coefficient, blade loading coefficient and the axial thrust.

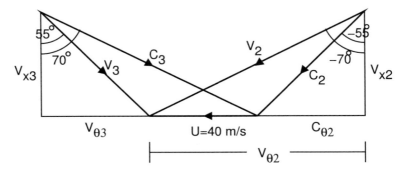

Fig. 3.11 Illustration for worked Example 3.2

Solution: From the given information, we can write $\beta_2 = \alpha_3 = 55°$ and $\beta_3 = \alpha_2 = 70°$ in the usual stage nomenclature. From the geometric construction in Fig. 3.11, it is clear that $C_2 = V_3$ and $C_3 = V_2$. Therefore,

$$h_2 - h_3 = \frac{C_3^2 - C_2^2}{2} = \frac{V_2^2 - V_3^2}{2}$$

and

$$h_{0,2} - h_{0,3} = \frac{V_2^2 - V_3^2}{2} - \frac{C_2^2 - C_3^2}{2} = V_2^2 - V_3^2.$$

Hence,

$$\mathcal{R} = \frac{h_2 - h_3}{h_{0,2} - h_{0,3}} = 0.5.$$

Note that the velocity diagram for the 50% reaction case is symmetrical about the vertical centerline. For values of $\mathcal{R}$ less 50%, the velocity diagram is skewed to the left approaching the diagram in Fig. 3.6 in the limit as $\mathcal{R} \to 0$. For values of $\mathcal{R}$ greater than 50%, the diagram is skewed to the right.

From the velocity triangle, it can be seen that $V_2 \cos 70 = C_2 \cos 55$ and $V_2 \sin 70 = C_2 \sin 55 + 40$.

This can be solved to give $V_2 = 88.65$ m/s and $C_2 = 52.86$ m/s.

It follows that $C_3 = 88.65$ m/s and $V_3 = 52.86$ m/s.

Also $V_{\theta,2} = V_2 \sin 70 = 83.3$ m/s, $V_{\theta,3} = V_3 \sin 55 = 43.3$ m/s and $V_{x,2} = V_{x,3} = 30.32$ m/s.

Flow coefficient, $\phi = V_{x,2}/U = (30.32)/(40) = 0.758$.

Blade loading coefficient, $\psi = (V_{\theta,2} - V_{\theta,3})/U = (83.3 + 43.3)/40 = 3.165$.

Power developed $= \dot{m} U (V_{\theta,2} - V_{\theta,3}) = (1)(40)(83.3 + 43.3) = 5064$ W.

Axial thrust is zero since there is no change in the axial velocity.

Example 3.3 A multi-stage axial compressor takes in air at 298 K and compresses it through a pressure ratio of 6:1. The mean blade speed is 250 m/s. The degree of

Fig. 3.12 Illustration for worked Example 3.3

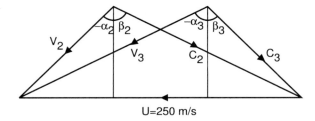

U=250 m/s

reaction, flow coefficient and blade loading coefficient for each stage are 0.5, 0.5 and 0.3. Determine the number of stages required as well as the blade and flow angles (Fig. 3.12).

Solution: From Eq. 3.9, we can get $\dot{W}/\dot{m} = \psi U^2$. Therefore,

$$\dot{W}/\dot{m} = (0.3)(250^2) = -18{,}750 \text{ J/kg}$$

per stage (keeping in mind that $\dot{W}$ is negative for a compressor).
From Eq. 2.17, we know that

$$\dot{W}/\dot{m} = h_{0,1} - h_{0,2} = C_p T_{0,1}\left(1 - \frac{T_{02}}{T_{01}}\right)$$

for the entire machine. If we assume the compression process to be isentropic, then

$$\frac{T_{0,2}}{T_{0,1}} = \left(\frac{P_{0,2}}{P_{0,1}}\right)^{\frac{\gamma-1}{\gamma}}.$$

Therefore,

$$\frac{\dot{W}}{\dot{m}} = C_p T_{0,1}\left[1 - \left(\frac{P_{0,2}}{P_{0,1}}\right)^{\frac{\gamma-1}{\gamma}}\right]$$

for the whole machine. With $C_p = 1005$ J/kg K and $\gamma = 7/5$ and $T_{0,1} = 298$ K and $P_{0,2}/P_{0,1} = 6$, we get

$$\dot{W}/\dot{m} = (1005)(298)(1 - 6^{2/7}) = -200.212 \text{ kJ/kg}$$

for the entire machine.
Hence, the number of stages required is $(-200{,}212)/(-18{,}750) \approx 11$.
From the velocity diagram, we can see that $V_{\theta,2} + C_{\theta,2} = U$. But $V_{\theta,2} = V_{x,2}\tan\alpha_2$ and $C_{\theta,2} = C_{x2}\tan\beta_2$. Also, $C_{x,2} = V_{x,2} = \phi U$. Therefore,

$$\phi(\tan\alpha_2 + \tan\beta_2) = 1. \tag{3.14}$$

From the definition of ψ, we get $V_{\theta,3} - V_{\theta,2} = \psi U$, where we have used the fact that $\dot{W}$ is negative for a compressor. If we substitute for $V_{\theta,3}$ and $V_{\theta,2}$, we get

$$\phi(\tan\alpha_3 - \tan\alpha_2) = \psi.$$

Since $\mathcal{R} = 1.\nabla$, $\alpha_3 = \beta_2$. Therefore,

$$\phi(\tan\beta_2 - \tan\alpha_2) = \psi. \tag{3.15}$$

Upon solving these two equations, we finally get $\alpha_2 = 35°$ and $\beta_2 = 52.4°$. Note that the comments given in the previous problem for the velocity diagram are applicable in this case also. The reader should also study the differences between the two diagrams carefully.

Acknowledgements I am grateful to Prof. Korpela of the Ohio State University for his help in formulating the exercise problems in this chapter.

Exercises

3.1 Fluid ($C_p = 1256$ J/kg K) enters the rotor in a single-stage axial flow turbine with an absolute velocity of fluid of 600 m/s and leaves with an absolute velocity of 300 m/s. The axial velocity remains constant and the blade speed is 300 m/s. The mass flow rate is 5 kg/s and $\alpha_1 = 65°$.

1. Construct the rotor inlet and exit velocity diagrams showing the axial and tangential components of the absolute velocities.
2. Evaluate the change in stagnation and static enthalpy across the rotor.
3. Evaluate the power developed by the moving blades.
4. Evaluate the average driving force exerted on the blades.
5. Evaluate the axial thrust.
6. Calculate the flow coefficient, the blade loading coefficient and the degree of reaction.

[115.04 kJ/kg, −19.96 kJ/kg, 575.21 kW, 1917.36 N, 0 N, 0.845, 1.278, −0.1734]

3.2 A small axial flow turbine is designed to produce 50 hp with a mass flow rate of 0.45 kg/s of air. The inlet stagnation temperature is 340 K. The rotor operates at 50,000 rpm and the mean blade diameter is 0.1 m. Evaluate

1. the average driving force on the turbine blades,
2. the change in the tangential component of the absolute velocity across the rotor and
3. the total pressure ratio across the turbine.

[140.64 N, 312.5 m/s, 2.6]

3.3 Consider a stage of a multi-stage axial turbine. The inlet fluid angle to the stator is $\alpha_1 = -37°$ and the outlet angle from the stator is $\alpha_2 = 60°$. For the rotor, $\beta_2 = 37°$ and $\beta_3 = -60°$. The blade speed is 150 m/s and the axial velocity remains constant. Determine

1. the axial velocity,
2. the work done by the fluid on the rotor blades,
3. the axial thrust and
4. the number of stages if the inlet stagnation temperature to the turbine is 950 K, the mass flow rate is 400 kg/s and the power output is 145 MW.

[153.2 m/s, 57.15 kJ/kg, 0 N, 7 stages]

3.4 An axial turbine has a total pressure ratio of 5 to 1, with an inlet total temperature of 800 K. The blade loading coefficient is 1.1 and the flow coefficient is 0.6. The absolute velocity at the inlet to the stator row is at angle 6° from the axial direction. Find

1. the angle at which the absolute velocity leaves the stator,
2. the angle of the relative velocity at the inlet of the rotor and
3. the angle at which the relative velocity leaves the rotor.

Draw the velocity diagrams at the inlet and outlet of the rotor.
[$-62.7°$, 15.2°, 57.4°]

3.5 Air enters the stator of an axial turbine stage axially. For the rotor, $\phi = 0.55$ and $\psi = 1.5$ and the axial velocity remains constant across the stage.

1. Draw the velocity diagrams for the stage.
2. Determine the angle at which relative velocity leaves the rotor.
3. Find the stator exit angle.
4. If the turbine has two stages, an inlet stagnation temperature of $T_{01} = 1450$ K and blade speed $U = 350$ m/s, find the stagnation temperature of the gas at the exit of the turbine and the stagnation pressure ratio for the turbine.
5. Assuming that the density ratio across the turbine based on static temperature and pressure ratios is the same as that based on the stagnation temperature and stagnation pressure ratios, find the ratio of the cross-sectional areas across the two-stage turbine.

[61.2°, $-70.5°$, 1084 K, 2.76, 2.07]

3.6 Consider the rotor of an axial compressor for which the blade speed is 150 m/s and the axial velocity is 75 m/s (and remains constant). Further, $C_{\theta,2} = 10$ m/s, $V_{\theta,1} = -35$ m/s, $T_{0,1} = 340$ K and $P_{0,1} = 185$ kPa.

1. Draw the velocity diagrams at the inlet and exit of the rotor.
2. Find the work done per unit mass flow through the compressor.
3. Draw the states in a T-s diagram.
4. Find the stagnation and static temperatures between the rotor and the stator.

5. Find the stagnation pressure between the rotor and the stator.

[29,250 J s/kg, 369.1 K, 353.6 K, 246 kPa]

3.7 In an axial flow air compressor, $\phi = 0.6$ and $\mathcal{R} = 1.\triangle$. At the inlet to the stator, $\alpha_1 = 60°$. Assuming the stage to be normal,

1. draw the velocity diagrams at the inlet and outlet of the rotor
2. determine the blade loading coefficient,
3. determine at what angle the relative velocity enters the rotor,
4. determine at what angle the relative velocity leaves the rotor and
5. determine at what angle the absolute velocity leaves the stator.

[2.077, 83.1°, 81.4°, −60°]

3.8 Air enters the rotor of a multi-stage compressor at $\alpha_1 = 40°$ and $\beta_1 = -60°$. The exit angle leaving the rotor is $\beta_2 = -40°$ and $\alpha_2 = 60°$. The mean radius of the rotor is 0.3 m. The axial velocity is constant and has a value $V_x = 125$ m/s. The inlet air is atmospheric at pressure 1 atm and temperature 20 °C. What should the shaft speed be under these conditions?
   [3555 rpm]

3.9 A single stage of a multi-stage axial compressor is shown in Fig. 3.10. The inlet air angle to the rotor is $\alpha_1 = 35°$ and the blade angle is $\beta_1 = -60°$. The outlet blade angle is $\beta_2 = -35°$. The gas inlet angle to the stator is $\alpha_2 = 60°$. The stage is normal and the axial velocity is constant through the compressor.

1. Why does the static pressure rise across both the rotor and the stator?
2. Draw the velocity triangles.
3. If the blade speed is $U = 290$ m/s, find the axial velocity.
4. Find the work done per unit mass flow for a stage and the increase in stagnation temperature across it.
5. The stagnation temperature at the inlet is 300 K. The overall stagnation pressure ratio is 17.5. Determine the number of stages in the compressor.

[281 m/s, 198.24 kJ s/kg, 197.3 K, 2]

3.10 Air at static pressure equal to 100 kPa and static temperature equal to 300 K enters an axial flow compressor with velocity 180 m/s. The tip diameter is 0.5 m, the hub diameter is 0.4 m and the shaft speed is 6000 rpm. Air enters the stage axially and leaves axially. The rotor turns the flow through 40°.

1. Draw the inlet and exit velocity diagrams for the rotor.
2. Draw the blade shapes.
3. Determine the mass flow rate.
4. Determine the power required.
5. What is the total pressure ratio for the stage?
6. Determine the flow coefficient and the blade loading factor.
7. Determine the degree of reaction.

[11.3 kg/s, 184.7 kW, 1.2, 0.98, 0.82, 1.4]

Chapter 4
Basics of Combustion Thermodynamics

In this chapter, some basic concepts in combustion are discussed. The combustion of a hydrocarbon fuel in a combustor is an important aspect of a gas turbine engine. Today, the most important issue relating to the combustor is emission and its control. There are also other issues such as combustion efficiency, stable operation and thermal stresses. Some of the basic ideas on fuels and combustion of the fuels discussed in this chapter will allow the reader to gain a better understanding of these issues. For a more detailed discussion of the topics related to combustion, the reader is referred to the book by Turns.

4.1 Enthalpy of Formation and Sensible Enthalpy

We have already defined in Sect. 2.4 the enthalpy per unit mass as $h = e + Pv$, where e is the internal energy. The connection between e and the modes of energy storage in molecules such as translation, rotation, and vibration was also discussed. It was also shown that the energy stored in these modes is dependent only on the temperature of the gas. However, the energy stored in the chemical bonds that make up the molecules was not accounted for in this development. This energy is somewhat like "dead storage" in a reservoir or dam and does not play a role so long as there are no chemical reactions. Once chemical reactions begin to take place, new product species are formed from the reactant species and it is important to account for the energy used to break the bonds in the reactant species and form new bonds in the product species.

  With this in mind, for any chemical species, the enthalpy at a given temperature T is defined to be the sum of the enthalpy of formation, $\bar{h}_f^\circ (T_{ref})$ and a sensible enthalpy, $\Delta \bar{h}(T)$. The overbar denotes that the enthalpies are on a molar basis. The enthalpy of formation is evaluated at the reference standard state which is at temperature

© The Author(s), under exclusive license to Springer Nature Switzerland AG 2022   43
V. Babu, *Fundamentals of Propulsion*,
https://doi.org/10.1007/978-3-030-79945-8_4

Table 4.1 Curve fit coefficients for $\bar{C}_p$ (gas phase)

Species	T (K)	a_1	a_2	a_3	a_4	a_5
CO_2	298–1200	24.99735	55.18696	−33.69137	7.948387	−0.136638
	1200–6000	58.16639	2.720074	−0.492289	0.038844	−6.447293
H_2O	500–1700	30.09200	6.832514	6.793435	−2.534480	0.082139
	1700–6000	41.96426	8.622053	−1.499780	0.098119	−11.15764
N_2	298–6000	26.09200	8.218801	−1.976141	0.159274	0.044434
O_2	298–6000	29.65900	6.137261	−1.186521	0.095780	−0.219663

Source http://webbook.nist.gov

$T_{ref} = 298\,\text{K}$ and pressure 1 atm. Subscript f refers to formation and superscript $^\circ$ refers to standard state. Thus,

$$\bar{h}(T) = \bar{h}_f^\circ + \Delta\bar{h}(T). \tag{4.1}$$

The sensible enthalpy depends only on the temperature and it is easy to see that the enthalpy that we have used so far is actually the sensible enthalpy. However, in combustion calculations, the sensible enthalpy has to be calculated from tabulated values for enthalpy, $\bar{h}$ or by using a curve fit polynomial for $\bar{C}_p$. Sensible enthalpy values for some gases encountered routinely in combustion are given in Table 4.3 at the end of the chapter.

Polynomial curve fit functions for $\bar{C}_p$ for various gases encountered in common combustion applications are given in Table 4.1. Here,

$$\bar{C}_P = a_1 + a_2\left(\frac{T}{1000}\right) + a_3\left(\frac{T}{1000}\right)^2 + a_4\left(\frac{T}{1000}\right)^3 + a_5\left(\frac{T}{1000}\right)^{-2}.$$

Thus, $\Delta\bar{h}(T) = \bar{C}_p(T - T_{ref})$.

The enthalpy of formation is defined as the heat required to form the species at the standard state from the basic elements which make up the species at the standard state. It is usually determined from experiments involving the combination of elements to form a particular species. When the species is not formed from the elements but from some other substances, then the heat required to form these substances from their constituent elements is considered to determine the enthalpy of formation of the particular species. The enthalpy of formation accounts for the energy contained in the chemical bonds between atoms that make up the molecules of the species over and above the energy of the bonds of the elements. By definition, enthalpies of formation of elements in their natural states at 298 K, like O_2, N_2 in gaseous state and C in solid state, are zero.

The potential energy due to forces between two atoms can be modeled with the well-known Lennard-Jones 6–12 potential,

Fig. 4.1 Lennard-Jones
6–12 potential curve

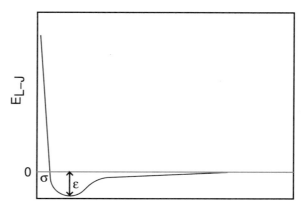

Interatomic separation distance

$$E_{L-J} = 4\epsilon \left[\left(\frac{\sigma}{r} \right)^{12} - \left(\frac{\sigma}{r} \right)^{6} \right].$$

Here, σ is the atomic radius, ϵ is the bond energy and r is the interatomic separation distance. Values of the Lennard-Jones parameters (σ and ϵ) for various species are available in *An Introduction to Combustion by Turns*. It should be noted that the force between the atoms is attractive for $r > \sigma$ and repulsive for $r < \sigma$. This is illustrated in Fig. 4.1. Hence, an amount of energy ϵ has to be spent to break a bond between two atoms or, an amount of energy ϵ is released when the bond between to atoms is formed. The bond energies for some typical bonds are given in Table 4.2.

It should be noted that bond energy is defined as the energy required to break the bond. Also, note that bond energies for the double bond are higher than those of the single bond and those for the triple bond are even higher. The bond energy constitutes the major part of the energy contained in the molecule. The enthalpy of formation can therefore be approximately estimated using bond energies. This is demonstrated in the following worked example.

Example 4.1 Determine the enthalpy of formation of gaseous methane, acetylene (C_2H_2), ethylene (C_2H_4), CO_2 and CO. Assume the enthalpy of formation of $C_{(g)}$ from the element C(s) to be 0.7184×10^6 kJ/kmol.

Solution: Gaseous methane contains four C−H bonds. Four H atoms are formed from the dissociation of two H_2 molecules. Carbon in the gaseous phase is formed from carbon in the solid state. Therefore, the enthalpy of formation of $CH_4 = -4 \times 0.413 \times 10^6 + 2 \times 0.436 \times 10^6 + 0.7184 \times 10^6$ kJ/kmol $= -61,600$ kJ/kmol. The actual value is $-74,850$ kJ/kmol.

Acetylene contains one C≡C and two C−H bonds. Two H atoms are formed from the dissociation of a H_2 molecule. Proceeding as before, the enthalpy of formation of $C_2H_2 = -2 \times 0.413 \times 10^6 - 0.814 \times 10^6 + 0.436 \times 10^6 + 2 \times 0.7184 \times$

Table 4.2 Bond energies (gas phase)

Bond	Energy (kJ/kmol)
H−H	0.436×10^6
C−C	0.348×10^6
C−H	0.413×10^6
O−H	0.456×10^6
C=O	0.804×10^6
C−O	0.36×10^6
C≡O	1.072×10^6
C≡C	0.814×10^6
O−O	0.139×10^6
$(O=O)_{O_2}$	0.498×10^6

10^6 kJ/kmol $= 232{,}800$ kJ/kmol, which is reasonably close to the actual value of 226,736 kJ/kmol.

Ethylene contains one C=C and four C−H bonds. Four H atoms are formed from the dissociation of two H_2 molecules. Therefore, the enthalpy of formation of $C_2H_4 = -4 \times 0.413 \times 10^6 - 0.615 \times 10^6 + 2 \times 0.436 \times 10^6 + 2 \times 0.7184 \times 10^6$ kJ/kmol $= 41{,}800$ kJ/kmol. The book value is 52,470 kJ/kmol.

CO_2 contains two C=O bonds. Two O atoms are formed from the dissociation of one O_2 molecule. Carbon in the gaseous phase is formed from carbon in the solid state. Hence, the enthalpy of formation of $CO_2 = -2 \times 0.804 \times 10^6 + 0.498 \times 10^6 + 0.7184 \times 10^6$ kJ/kmol $= -391{,}600$ kJ/kmol. The actual value is $-393{,}510$ kJ/kmol.

In CO, the carbon atom is joined to the oxygen atom by a triple bond. Hence, the enthalpy of formation of CO $= (-2 \times 1.072 \times 10^6 + 0.498 \times 10^6 + 2 \times 0.7184 \times 10^6)/2$ kJ/kmol $= -104{,}600$ kJ/kmol. The actual value is $-110{,}530$ kJ/kmol.

Note that the enthalpies of formation of the species considered above are large and negative. The values are negative since heat is released during the formation of the species from its constituent elements.

In the case of a molecule such as benzene, energy due to resonance also has to be accounted for. This arises since benzene can resonate between five distinct molecular structures without changing the number or nature of the bonds between the atoms. The interested reader can consult any advanced combustion book for details.

4.2 Properties of Ideal Gas Mixtures

Consider a gas mixture composed of m_i kg each of a number of components. The number of moles of species i is $n_i = m_i/M_i$, where M_i is the molecular weight. The total number of moles in the mixture is then $n = \sum_i n_i$. Mole fraction of species i is given as $y_i = n_i/n$ and mass fraction as $x_i = m_i/m$.[1] Here, $m = \sum_i m_i$ is the total mass of the mixture.

The mixture molecular weight M is given as $M = m/n$. From the definitions of the mole fraction and mass fraction given above, it is easy to show that

$$M = \sum_i y_i M_i = \frac{1}{\sum_i x_i/M_i}.$$

Equation of state for a mixture is defined in the same manner as that of a single component gas, with the particular gas constant being evaluated using the mixture molecular weight. Thus, $PV = mRT = n\mathcal{R}T$, where $R = \mathcal{R}/M$ and $\mathcal{R}$ is the universal gas constant.

Two new concepts, namely partial pressure and partial volume are commonly used when dealing with a mixture of gases. Dalton's model of a mixture of gases assumes each component i to be at its own partial pressure, P_i but occupying the same volume as that of the mixture and at the same temperature. In Amagat's model, each component occupies a partial volume V_i, but has the same pressure and temperature as that of the mixture. Obviously, $\sum_i P_i = P$ and $\sum_i V_i = V$. In Dalton's model, $P_i V = m_i R_i T$, where $R_i = \mathcal{R}/M_i$. Since $PV = mRT$ for the mixture, it follows that

$$P_i = \frac{m_i}{m} \frac{R_i}{R} = x_i \frac{M}{M_i}.$$

Alternatively, $P_i V = n_i \mathcal{R}T$ and since $PV = n\mathcal{R}T$, it follows that $P_i = y_i$.

In Amagat's model, $PV_i = m_i R_i T$ and it is easy to show that $V_i = x_i (M/M_i) = y_i$.

For combustion calculations, air can be assumed to contain 21% O_2 and 79% N_2 by volume. Thus, every kmole of air contains 0.21 kmoles of O_2 and 0.79 kmoles of N_2 and so $y_{O_2} = 0.21$ and $y_{N_2} = 0.79$. Hence, every kmole of O_2 is accompanied by $0.79/0.21 \approx 3.76$ kmoles of N_2. This means that for every kmole of O_2 required for combustion, $1/0.21 \approx 4.76$ moles of air has to be supplied.

Properties of mixtures can be calculated in a straightforward manner as follows. The specific enthalpy of the mixture (on a mass basis) is given as

$$h = \frac{\sum_i m_i h_i}{\sum_i m_i} = \sum_i x_i h_i.$$

Specific enthalpy on a molar basis is given as

$$\bar{h} = \frac{\sum_i n_i \bar{h}_i}{\sum_i n_i} = \sum_i y_i \bar{h}_i.$$

---

[1] This notation is customarily used in Thermodynamics text books. Combustion text books, on the other hand, use x_i to denote mole fraction and y_i to denote mass fraction.

Each $\bar{h}_i$ in this expression has to be evaluated using Eq. 4.1. Any other property of the mixture can be calculated in a similar fashion.[2]

4.3   Combustion Stoichiometry

Consider the complete combustion of octane (C_8H_{18}) with the theoretical (stoichiometric) amount of air. Here, complete combustion of a hydrocarbon fuel with air implies that the combustion products are fully oxidized, i.e., only CO_2 and H_2O are the only products. N_2 is assumed to be inert. The combustion equation can be written as

$$C_8H_{18} + 12.5(O_2 + 3.76N_2) \rightarrow 8CO_2 + 9H_2O + (12.5)(3.76)N_2.$$

Thus, each kmole of fuel requires 12.5 kmoles of O_2 or $(12.5)(4.76) = 59.5$ kmoles of air for complete combustion. Therefore, the stoichiometric air–fuel ratio on a molar basis is 59.5 kmole of air per kmole of fuel. Air–fuel ratio on a mass basis is

$$= 59.5 \frac{\text{kmole of air}}{\text{kmole of fuel}} \times 28.97 \frac{\text{kg}}{\text{kmole of air}} \times \frac{1}{114} \frac{\text{kmole of fuel}}{\text{kg}}$$
$$= 15 \frac{\text{kg of air}}{\text{kg of fuel}}.$$

For any hydrocarbon fuel with composition of the form C_nH_{2n+2}, the complete combustion equation can be written as

$$C_nH_{2n} + 2\left(\frac{3n+1}{2}\right)(O_2 + 3.76N_2) \rightarrow n\,CO_2 + (n+1)H_2O$$
$$+ \left(\frac{3n+1}{2}\right)(3.76)N_2.$$

Air–fuel ratio on a molar basis is

$$= \left(\frac{3n+1}{2}\right)(4.76)\frac{\text{kmole of air}}{\text{kmole of fuel}}.$$

On a mass basis, air–fuel ratio

---

[2] Ratio of specific heats for the mixture, $\gamma = (\sum_i x_i C_{p,i})/(\sum_i x_i C_{v,i})$.

$$= \left(\frac{3n+1}{2}\right)(4.76)\frac{\text{kmole of air}}{\text{kmole of fuel}} \times 28.97\frac{\text{kg}}{\text{kmole of air}}$$

$$\times \frac{1}{12n+2n+2}\frac{\text{kmole of fuel}}{\text{kg}}.$$

For large values of n, this expression reduces to $\frac{3n}{2}(4.76)(28.97)\frac{1}{14n}$, which is equal to $14.77\frac{\text{kg of air}}{\text{kg of fuel}}$.

This is a remarkable result since it is independent of n. Although this result has been obtained for large n, it can be used as a good engineering approximation even for small values of n. It is left as an exercise to the reader to derive this for other families of hydrocarbon fuels such as C_nH_{2n} and C_nH_{2n-2}.

In many practical applications, the amount of air used is in excess of the stoichiometric amount of air for several reasons. The most important reason is that the presence of the additional amount of O_2 and N_2 serves to dilute the combustion products thereby reducing the temperature to manageable levels. Also, the presence of excess O_2 tends to stabilize the combustion and improves the combustion efficiency, up to a certain extent. There is also an attendant effect on the composition of the product gas mixture. This idea of excess air leads naturally to the definition of the equivalence ratio,

$$\Phi = \frac{\text{Actual Fuel-air ratio}}{\text{Stoichiometric Fuel-air ratio}}.$$

Fuel–air ratio is the reciprocal of the air–fuel ratio. An equivalence ratio of unity represents a stoichiometric mixture. Equivalence ratios less than one and greater than one represent a fuel-lean mixture and fuel-rich mixture, respectively.

Example 4.2 A dry analysis of the products from the combustion of methane with air is as follows:
$CO_2 = 9\%$, $CO = 2\%$, $O_2 = 2.11\%$ and $N_2 = 86.89\%$ by volume.
Calculate the equivalence ratio.

Solution: Since the analysis is on a volumetric basis, it is the same as the molar basis. Also, note the combustion of methane would have produced water vapor, but the analysis is on a dry basis. If we assume 100 kmole of products, a kmole of fuel and b kmole of air, then the combustion equation can be written as

$$aCH_4 + b(O_2 + 3.76N_2) \rightarrow 9CO_2 + (2a)H_2O + (86.89)N_2$$
$$+ 2CO + (2.11)O_2.$$

Carbon atom balance gives $a = 11$ and N atom balance gives $b = 23.11$. Oxygen atom balance is automatically satisfied. Hence, the actual air–fuel ratio is $(4.76)(23.11)/(11) = 10$ kmole of air/kmole of fuel. For complete combustion,

$$CH_4 + 2(O_2 + 3.76N_2) \rightarrow CO_2 + 2H_2O + (2)(3.76)N_2.$$

The stoichiometric air–fuel ratio is thus equal to $2 \times 4.76 = 9.52$ kmole of air/kmole of fuel. Hence, the equivalence ratio is 1.05.

4.4   Enthalpy of Combustion and Calorific Value

Consider a combustor operating at a steady state in which the reactants enter at a temperature T_1. Combustion takes place and after removal of heat, the products leave at the same temperature T_1. If we apply Eq. 2.17 to the combustor, we get

$$q = -(h_{\text{reactants}} - h_{\text{products}}).$$

Here, we have ignored the change in kinetic energy and $w = 0$ since there is no work interaction. This equation can also be written as

$$Q = -(H_{\text{reactants}} - H_{\text{products}}), \qquad (4.2)$$

where Q has units of J. For an exothermic reaction, the enthalpy of the reactants is greater than that of the products and hence heat is released in the combustion reaction. This is illustrated graphically in Fig. 4.2.

The enthalpy of reaction (or combustion) ΔH_R is defined as $\Delta H_R = H_{\text{products}} - H_{\text{reactants}}$. For combustion reactions that are exothermic, ΔH_R is negative. This is because the heat released in the chemical reactions has to be *removed* to maintain the products at the same temperature as the reactants. ΔH_R is usually evaluated at the reference temperature (298 K). In combustion applications, since the reactants are fuel plus air, the calorific value of the fuel is numerically equal to ΔH_R. However,

Fig. 4.2 Illustration of enthalpy of combustion

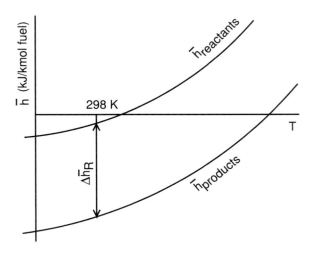

here a distinction needs to be made depending upon whether the water in the product stream is in vapor or liquid form. The resulting value of ΔH_R is called the lower calorific value (LCV) in the former case and higher calorific value (HCV) in the latter case (since the latent heat of vaporization is also removed).

Example 4.3 Determine the enthalpy of combustion of methane at 298 K. Assume the enthalpy of formation of methane to be $-74,850$ kJ/kmol.

Solution: Combustion of methane with a stoichiometric amount of air is represented by the equation:

$$CH_4 + 2(O_2 + 3.76N_2) \rightarrow CO_2 + 2H_2O_{(v)} + (2)(3.76)N_2.$$

Thus,

$$\bar{H}_{reactants} = \sum_i n_i h_i = \sum_i n_i \left(\bar{h}^\circ_{f,i} + \Delta h_i\right) = \sum_i n_i \bar{h}^\circ_{f,i},$$

where the last equality follows from the fact that we are evaluating the enthalpy of combustion at 298 K. Thus, $\bar{H}_{reactants} = -74,850$ kJ. Similarly, using the values from Table 4.3, we can get $\bar{H}_{products} = -877,180$ kJ. Therefore,

$$\Delta \bar{H}_R = -877,180 - (-74,850) = -802,330 \text{ kJ/kmol of fuel.}$$

On a mass basis, $\Delta H_R = -802,330$ (kJ/kmol of fuel)/16 (kg/kmol of fuel) $= -50.14$ MJ/kg of fuel. The lower calorific value of the fuel is thus 50.144 MJ/kg of fuel.

4.5   Adiabatic Flame Temperature

If the combustor considered in the previous subsection is insulated, then the energy equation (after neglecting changes in kinetic energy) reduces to

$$-(h_{reactants} - h_{products}) = 0$$

or

$$H_{reactants} = H_{products}. \tag{4.3}$$

This implies that the products leave at a temperature at which the enthalpy of the products is equal to the enthalpy of the incoming reactants at temperature T_1. The products in this case will leave at a temperature that is different from that of the reactants. This is illustrated graphically in Fig. 4.3. The temperature of the products in this case is called the adiabatic flame temperature. This is an important concept since it represents the theoretical maximum product temperature that can be realized at the combustor exit.

Table 4.3 Sensible enthalpies and enthalpy of formation of some gases

T (K)	$\Delta \bar{h}(T)$			
	O_2 (MJ/kmol)	N_2 (MJ/kmol)	CO_2 (MJ/kmol)	H_2O (MJ/kmol)
	$\bar{h}_f^\circ = 0$	$\bar{h}_f^\circ = 0$	$\bar{h}_f^\circ = -393.52$	$\bar{h}_f^\circ = -241.83$
298	0.0	0.0	0.0	0
300	0.03	0.04	0.07	0.062
400	3.02	2.95	4	3.458
500	6.13	5.92	8.31	6.92
600	9.32	8.93	12.91	10.5
700	12.59	12.01	17.75	14.19
800	15.91	15.14	22.81	18
900	19.29	18.32	28.03	21.94
1000	22.71	21.55	33.4	26
1100	26.18	24.83	38.89	30.19
1200	29.70	28.15	44.47	34.51
1300	33.25	31.51	50.15	38.94
1400	36.84	34.91	55.89	43.49
1500	40.46	38.34	61.71	48.15
1600	44.12	41.81	67.57	52.91
1700	47.81	45.31	73.48	57.76
1800	51.53	48.84	79.44	62.69
1900	55.28	52.4	85.42	67.7
2000	59.06	55.98	91.44	72.79
2100	62.86	59.58	97.49	77.94
2200	66.69	63.2	103.6	83.16
2300	70.54	66.84	109.7	88.42
2400	74.41	70.5	115.8	93.74
2500	78.31	74.17	121.9	99.11
2600	82.22	77.85	128.1	104.5
2700	86.16	81.55	134.2	110
2800	90.11	85.26	140.4	115.5
2900	94.08	88.97	146.6	121
3000	98.07	92.7	152.8	126.5
3100	102.1	96.43	159.1	132.1
3200	106.1	100.2	165.3	137.8
3300	110.1	103.9	171.6	143.4
3400	114.2	107.6	177.8	149.1
3500	118.2	111.4	184.1	154.8
3600	122.3	115.1	190.4	160.5
3700	126.4	118.9	196.7	166.2
3800	130.5	122.6	203	172

(continued)

Table 4.3 (continued)

T (K)	$\Delta \bar{h}(T)$			
	O_2 (MJ/kmol) $\bar{h}^\circ_f = 0$	N_2 (MJ/kmol) $\bar{h}^\circ_f = 0$	CO_2 (MJ/kmol) $\bar{h}^\circ_f = -393.52$	H_2O (MJ/kmol) $\bar{h}^\circ_f = -241.83$
3900	134.6	126.4	209.3	177.8
4000	138.7	130.2	215.6	183.6
4100	142.9	133.9	222	189.4
4200	147	137.7	228.3	195.2
4300	151.2	141.4	234.6	201
4400	155.4	145.2	241	206.9
4500	159.6	148.9	247.4	212.8
4600	163.8	152.7	253.7	218.7
4700	168	156.5	260.1	224.5
4800	172.2	160.2	266.5	230.5
4900	176.5	164	272.9	236.4
5000	180.7	167.7	279.3	242.3
5100	185	171.5	285.7	248.3
5200	189.3	175.3	292.1	254.2
5300	193.6	179.1	298.5	260.2
5400	197.9	182.9	305	266.2
5500	202.2	186.7	311.4	272.2
5600	206.6	190.5	317.9	278.2
5700	211	194.3	324.3	284.2
5800	215.4	198.1	330.8	290.2
5900	219.8	202	337.3	296.2
6000	224.2	205.8	343.8	302.3

Source http://webbook.nist.gov

Example 4.4 A stoichiometric mixture of methane and air at 298 K enters an insulated combustor. Determine the temperature of the products leaving the combustor.

Solution: We already know from the previous example that $\bar{H}_{\text{reactants}} = -74{,}850$ kJ. Now,

$$\bar{H}_{\text{products}} = \sum_i n_i h_i = \sum_i n_i \left(\bar{h}^\circ_{f,i} + \Delta h_i \right).$$

Upon expanding the right hand side, we get

$$\bar{H}_{\text{products}} = \left(\bar{h}^\circ_{f,CO_2} + 2\bar{h}^\circ_{f,H_2O} + 7.52\bar{h}^\circ_{f,N_2} \right)$$
$$+ \left(\Delta\bar{h}_{CO_2} + 2\Delta\bar{h}_{H_2O} + 7.52\Delta\bar{h}_{N_2} \right).$$

If we substitute values for the enthalpies of formation, we get

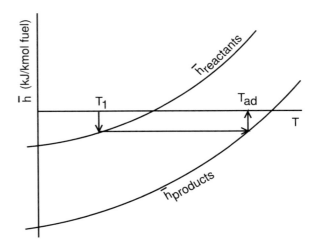

Fig. 4.3 Illustration of the adiabatic flame temperature concept

$$\bar{H}_{\text{products}} = -877,160 + \left(\Delta\bar{h}_{\text{CO}_2} + 2\Delta\bar{h}_{\text{H}_2\text{O}} + 7.52\Delta\bar{h}_{\text{N}_2}\right).$$

Equating this to $\bar{H}_{\text{reactants}}$ and rearranging the resulting expression give

$$\Delta\bar{h}_{\text{CO}_2} + 2\Delta\bar{h}_{\text{H}_2\text{O}} + 7.52\Delta\bar{h}_{\text{N}_2} = 802,330.$$

The temperature of the products has to be determined iteratively. It is reasonable to expect the value to lie between 298 and 5000 K. Hence, we start with an initial guess equal to $(298 + 5000)/2 \approx 2650$ K. Proceeding with the nearest entry in Table 4.3 which is 2600 K, the left hand side of the above equation comes out to be 955,394 kJ/kmol. This suggests that our initial guess is on the higher side. The next guess can be taken to be $(298 + 2600)/2 \approx 1449$ K. Let us take 1500 K as our next guess. The left hand side now comes out to be 446,326, which is less than the right hand side. The next guess is $(1500 + 2600)/2 = 2050$ K. Proceeding with 2100 K as the next guess, the left hand side is 701,411. Our next guess is $(2100 + 2600)/2 = 2350$ K. The nearest entry in Table 4.3 is 2400 K. This gives 833,440 for the left hand side. If we take 2300 K as our next guess, we get 789,176 for the left hand side. With linear interpolation, we can finally estimate 2330 K as the temperature of the products.

In this example, we have assumed CO_2, H_2O and N_2 to be the only products which is not realistic. At such high temperatures, CO_2 will decompose to form CO and H_2O will decompose to form OH and H radicals and N_2 combines with these radicals to form oxides of nitrogen such as NO. This is discussed in detail in the next chapter.

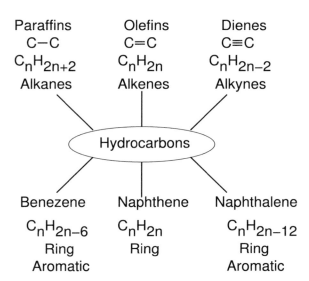

Fig. 4.4 Classification of hydrocarbon compounds

4.6 Fuels and Their Properties

In this section, different types of hydrocarbon fuels and some of their important properties are discussed. Hydrocarbon compounds can be broadly classified into two types, namely straight chain and ring (Fig. 4.4), based on the arrangement of the carbon atoms.

Straight-chain hydrocarbons are widely used as fuels. These can be classified into three types as paraffins, olefins and diolefins (or dienes) depending upon the bond between the carbon atoms. As shown in Fig. 4.4, paraffins have a $C-C$ single bond, olefins have a $C=C$ double bond and diolefins have a $C\equiv C$ triple bond. Ethylene (C_2H_4) is the first member of the olefin family and acetylene (C_2H_2) is the first member of the diolefin family. Paraffins contain the maximum possible number of H atoms for each C atom on account of the $C-H$ single bond and are said to be *saturated*. Olefins and diolefins, on the other hand, are unsaturated hydrocarbons. Since double and triple bonds have higher energy than single bonds, olefins and diolefins have higher calorific value than paraffins. Furthermore, as the ratio of carbon to hydrogen atoms is higher than that of paraffins, the combustion of olefins and diolefins produces more soot.

The paraffin family starts with methane (CH_4) and goes up to lubricating oils ($C_{50}H_{102}$) and beyond. Under standard reference conditions (298 K and 1 atm), the first few members of the paraffin family, namely methane, ethane, propane and butane, are gases.[3] Octane (C_8H_{18}) is a liquid but it evaporates rapidly when exposed

---

[3] Replacement of a H atom with an OH radical in a member of the paraffin family leads to the corresponding member of the alcohol family. For example, replacing a H in methane, CH_4, with OH results in methyl alcohol (methanol), CH_3OH.

to the atmosphere. Kerosene (n-dodecane, $C_{12}H_{26}$) is also a liquid but does not evaporate as rapidly, owing to its higher molecular weight. Furnace oil ($C_{20}H_{42}$) is probably the last member of the paraffin family that can be used as a fuel. Liquid hydrocarbons in the order of decreasing volatility are gasoline, naptha, kerosene, diesel oil and fuel oil. Paraffins with more than 20 carbon atoms are used as lubricating oils. With an increasing number of C atoms, the vapor pressure decreases and the viscosity increases.

Benzene, naphthene and naphthalene have a ring structure. The carbon atoms are joined by single bonds in naphthene, single and double bonds in benzene and double and triple bonds in naphthalene. The ring structured benzene and naphthalene exude a strong aroma (naphthalene balls or mothballs used in households is a good example). The complex ring structure with double and triple bonds are more difficult to break compared to straight-chain structure with single bonds. These are widely used, among other things, as solvents in industrial applications and are carcinogenic. Although inflammable, these are not safe (from a handling or emissions perspective) to be used as fuels.

Pure kerosene, which is actually a combination of several hydrocarbons, is called aviation kerosene and is used for aircraft propulsion. Kerosene mixed with gasoline is known as a jet mix product (JP) and has superior combustion properties. Blending of kerosene is also done with naptha, and other substances (additives) are added to the blended mixture to give a lower freezing point, higher density, lower sulfur content and suitable flash points as may be desired for the particular aircraft application. Additives are also introduced to achieve better storage and handling properties, such as increased corrosion resistance, reduced flammability to electrostatic discharge and reduced ice, gum and deposit formation.

Commercial aircraft use Jet A or Jet A-1 fuel. Both these fuels are kerosene based and are very similar except that Jet A has a freezing point of $-40\,°C$, while Jet A-1 has a freezing point of $-47\,°C$. Fuels intended for use in military applications need to satisfy more stringent requirements. A blend of kerosene and naptha with additives is known as Jet B fuel. Depending on the additives, this is further categorized into JP2, JP3 and JP4. JP4 has both paraffinic and aromatic content. JP5 has a higher flash point. JP6 and JP7 have higher thermal stability. JP8 has the highest heat release per unit mass, whereas JP9 has the highest heat release per unit volume and is a blend of three hydrocarbons. JP10 is a single chemical compound *exo-tetrahydrodicyclopentadiene*, $C_{10}H_{16}$.

4.6.1 Combustion Aspects of Fuels

Laminar flame speed in a fuel can be thought of as the speed with which a flame front will travel through a quiescent mixture of reactants. Conversely, if the reactant mixtures were moving with the laminar flame speed, then the flame front would be stationary. The latter perspective is relevant in aircraft propulsion since the fuel and air flow continuously through the combustion chamber. Hence, if the speed of the

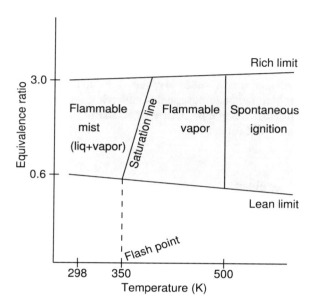

Fig. 4.5 Flammability limits of a kerosene type of fuel at 1 atm. Adapted from Elements of Gas Turbine Propulsion by Mattingly

fuel–air mixture is high when compared with the laminar flame speed, the flame is blown out. On the other hand, if the speed of the fuel–air mixture is too low, then the flame tries to flash back and is extinguished. The laminar flame speed is proportional to the square root of the thermal diffusivity (among other properties) of the reactant mixture. The thermal diffusivity and hence the laminar flame speed decrease with increasing molecular weight of the fuel. Once the number of carbon atoms in the fuel increases beyond 6, the laminar flame speed asymptotes to about 40 cm/s.

Flammability is an important characteristic of any fuel. It is basically the ability of a fuel to burn with a given amount of air at a given pressure and temperature. Contrary to what one might expect, a fuel does not always burn readily or in a stable manner. Combustion cannot be initiated or sustained if there is either too much air (lean limit) or too much fuel (rich limit). The flammability limits of a kerosene type of fuel at 1 atm pressure are shown in Fig. 4.5. For equivalence ratios below 0.5 or above 3, combustion is not at all possible at normal room temperature. This range is typical of all hydrocarbon fuels.[4] Hydrogen alone can burn over a wider range of equivalence ratios (0.25–6) even at room temperature and pressure. As shown in Fig. 4.5, under normal room conditions, kerosene exists as a liquid. As the temperature is increased, it becomes a two-phase mixture of liquid and vapor

---

[4] During actual operation, it may not always be possible to maintain the equivalence ratio everywhere in the combustor within this limit. To alleviate this problem, the equivalence ratio in small regions (called flame stabilization regions) inside the combustor is maintained within the allowed range. These regions are usually low-speed recirculation regions in which hot combustion products are available and thus they provide a high-temperature source to sustain combustion. Mechanical devices that are usually used to create these regions are called flame stabilizers (or more popularly, V-gutters, owing to their shape).

(mist). Once the saturation temperature is crossed, only the vapor phase is present. If the temperature is increased further, the vapor can ignite in the presence of air upon reaching the self ignition temperature, provided the equivalence ratio is within the lean and rich limit. If the temperature is less than the self ignition temperature, then an ignition source is required. From a perspective of fuel handling, higher self ignition temperatures are desirable. Self ignition temperature of paraffins decreases with increasing molecular weight, starting with 767 K for propane, 491 K for octane and 478 K for cetane.

Exercises

4.1 Determine the air–fuel ratio on a molar basis as well as the mass basis for the stoichiometric combustion of liquid n-dodecane ($C_{12}H_{26}$).
[88:1, 15:1]

4.2 An aviation gas turbine engine utilizing liquid n-dodecane as the fuel operates with an air–fuel ratio of 75:1 (mass basis). Determine the corresponding equivalence ratio.
[0.2]

4.3 Determine the enthalpy of combustion of liquid n-dodecane at 298 K. Take $\bar{h}^o_{f,C_{12}H_{26}} = -292,162$ kJ/kmol.
[44.55 MJ/kg]

4.4 Fuel (liquid n-dodecane) at 298 K enters the combustor in a gas turbine. Air enters the compressor at 100 kPa and 298 K. The compressor pressure ratio is 18 and the compression process is isentropic. The air–fuel ratio on a mass basis is 75:1. Assume complete combustion and negligible heat loss from the combustor. Determine the temperature at the exit of the combustor.
[1179 K]

4.5 If the exit temperature in the above problem can be 150% of the value that you obtained, what should be the air–fuel ratio on a mass basis?
[31.42:1]

4.6 Fuel (liquid n-dodecane) and air at 298 K enter a gas turbine. The air–fuel ratio on a mass basis is 60:1. The products of combustion leave the gas turbine at 1000 K and the specific fuel consumption is 0.25 kg/s of fuel per megawatt of output power. Determine the heat transfer rate from the combustor.
[620 kJ/s]

4.7 LPG (liquefied petroleum gas) is a mixture of 40% propane (C_3H_8) and 60% butane (C_4H_{10}) by volume. Determine its lower calorific value. Take $\bar{h}^o_{f,C_3H_8} = -103,847$ kJ/kmol and $\bar{h}^o_{f,C_4H_{10}} = -124,733$ kJ/kmol.
[46.04 MJ/kg]

4.8 Determine the adiabatic flame temperature when LPG is burnt with 150% of stoichiometric air. Assume that the reactants are at 298 K.
[1835 K]

4.9 Determine the stoichiometric air–fuel ratio on molar and mass basis for the combustion of ethanol (C_2H_5OH) with air. If ethanol is burnt with 200% stoichiometric air, what would be the result of a dry analysis of the combustion products? Determine the lower calorific value of ethanol assuming $\bar{h}^o_{f,C_2H_5OH(l)} = -277,690$ kJ/kmol.
[7.25% CO_2, 10.8% O_2, 81.95% N_2, 26.84 MJ/kg]

4.10 Ethane, ethylene and acetylene are, respectively, the members of the paraffin, olefin and diolefin families corresponding to $n = 2$. Determine the lower calorific value of the fuels and relate the values to their chemical structure. Take $\bar{h}^o_{f,C_2H_6} = -84,680$ kJ/kmol, $\bar{h}^o_{f,C_2H_4} = 52,280$ kJ/kmol and $\bar{h}^o_{f,C_2H_2} = 226,730$ kJ/kmol.
[47.6 MJ/kg, 47.25 MJ/kg, 48.29 MJ/kg]

Chapter 5
The Turbojet Engine

In this chapter, the operation of a basic turbojet engine is described. This is followed by a discussion of the individual components, namely, the compressor, combustor, turbine and the nozzle. This discussion brings out the details of the operation of these components along with important issues related to their operation. A thermodynamic analysis of the operation of the whole engine as well as the components is deferred till Chap. 7.

A longitudinal section of a single shaft turbojet engine is shown in Fig. 5.1. Four main components of the engine, namely, compressor, combustor, turbine and nozzle are also shown in this figure. An inlet is not explicitly shown in this figure but is usually present. The inlet is not a very elaborate design in the case of subsonic flight Mach numbers but is just a short diverging passage with a rounded leading edge preceding the compressor. In the case of supersonic flight Mach numbers, the inlet is a crucial part of the engine since the compression of the air is achieved entirely by deceleration of the air in the inlet. Supersonic intakes are discussed in detail in the last chapter.

The incoming air is compressed to pressures of the order of 30–40 atm in the compressor. Heat is added in the combustion chamber by burning the fuel (usually a kerosene fuel). This increases the stagnation temperature of the fluid. The work required to drive the compressor and other auxiliaries is extracted in the turbine. The fluid is then expanded in the nozzle to high velocities and the high velocity jet generates thrust as it leaves the nozzle. This, in a nutshell, is the description of a turbojet engine. A closer look on a component by component basis is given next.

V. Babu, *Fundamentals of Propulsion*,
https://doi.org/10.1007/978-3-030-79945-8_5

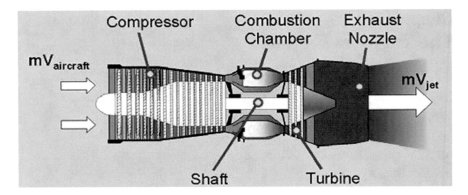

Fig. 5.1 Longitudinal section of a turbojet engine. *Courtesy* Rolls-Royce plc

5.1   Compressor

5.1.1   Nature of Compression Process

Equation 3.5 represents the relation between the change in the thermodynamic and the fluid dynamic properties, as the fluid flows through the compressor,

$$h_1 - h_2 = \frac{U_1^2 - U_2^2}{2} - \frac{C_1^2 - C_2^2}{2}.$$

This can be written in a differential form as follows:

$$dh = d\left(\frac{U^2}{2}\right) - d\left(\frac{C^2}{2}\right). \tag{5.1}$$

Since $h = e + Pv$, we can write $dh = de + Pdv + vdP$, and from Eq. 2.12, we can also write $de = \delta q_{\text{rev}} - Pdv$. Upon substituting $\delta q_{\text{rev}} = Tds$, we finally get $dh = Tds + vdP$. The above equation thus becomes

$$Tds + vdP = d\left(\frac{U^2}{2}\right) - d\left(\frac{C^2}{2}\right).$$

If we assume the compression process to be isentropic, then $ds = 0$. Thus

$$\frac{dP}{\rho} = d\left(\frac{r^2\omega^2}{2}\right) - d\left(\frac{C^2}{2}\right), \tag{5.2}$$

where we have used $v = 1/\rho$ and $U = r\omega$ with ω being the angular velocity of the rotor. This equation offers a nice insight into the nature of the compression

Fig. 5.2 Schematic illustration of the variation of pressure and temperature across a jet engine

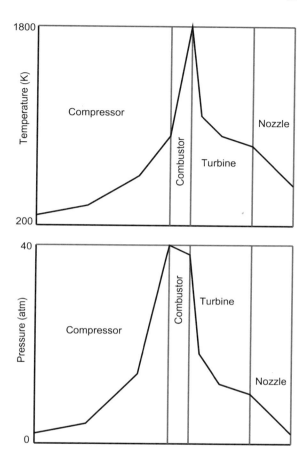

process. It clearly brings out the fact that the pressure rise along a streamline is due to a combination of two effects, namely, centrifugal and diffusion effect. The first term shows that, as the fluid moves radially outward, the pressure increases.[1] The second term shows that, as the fluid is decelerated in the streamwise direction in the compressor blade passage, pressure increases. The negative sign in front of the term accounts for the deceleration and the relative velocity (not the absolute velocity) is the streamwise velocity component inside the blade passage.

In a centrifugal compressor, the pressure rise is almost entirely due to centrifugal effect and so the second term is negligible ($dC \approx 0$). In an axial flow compressor, the streamlines are almost axial and so the first term is negligible ($dr \approx 0$) and the pressure rise is entirely from diffusion or deceleration of the flow in the streamwise direction. As already alluded to in Chap. 3, the combination of deceleration of the

---

[1] Conversely, this equation shows why the flow has to be radially outward in a centrifugal pump or compressor, where $dP > 0$ and radially inward in radial flow turbines, where $dP < 0$.

main flow and adverse pressure gradient in this case can cause the boundary layer on the blade surfaces to separate if the pressure rise is large enough. Consequently, the blade loading factor and the pressure rise that can be realized is limited. This is evident in Fig. 5.1, where the compressor has many more stages than the turbine, for the same work interaction. Typical values for pressure ratio across a single stage are of the order of 1.1 to 1.2. In contrast, the pressure rise across a centrifugal compressor can be as high as 4 or 5. However, the frontal area for a given pressure ratio is high for a centrifugal compressor. In addition, mutistaging of centrifugal compressors is difficult owing to the complex ducting required to redirect the radial outflow from one stage to an axial inflow for the next stage. Axial flow compressors are compact and can be readily multistaged. However, a large number of stages are required for achieving large overall pressure ratios. For instance, to achieve an overall pressure ratio of 30, about 15 stages, each with a pressure ratio of 1.1, are required.[2] Despite this, axial compressors are now exclusively used in aircraft jet engines.[3]

5.1.2   Design and Off-Design Operation

Figure 5.2 schematically shows the variation of pressure and temperature across a jet engine. Different segments in the pressure profile are shown to denote low, intermediate and high pressure stages. The pressure profile in the compressor shows the rise to follow a geometric progression, i.e., the rise is modest in the beginning (low pressure stages) and becomes steeper (intermediate and high pressure stages). As mentioned in Chap. 3, the height of the blades decreases to accommodate the increasing density of the fluid. This can be seen in Fig. 5.1. The decrease in the height of the blades is very small for the low pressure stages, since the pressure rise is not very high and the hub diameter and the casing diameter are almost constant. Beyond this, the blade height decreases rapidly with increasing pressure rise. Consequently, the casing diameter decreases continuously across the intermediate and high pressure stages, while the hub diameter increases across the intermediate stages and remains constant across the high pressure stages. This strategy is utilized to achieve reasonably constant axial velocity of the fluid from the first stage of the compressor to the last.

The area of the blade passage and the blade shape are designed for a particular mass flow rate (and hence a particular value of density and axial velocity) and blade speed. The velocity triangles for operation at design condition for one compressor stage are shown in Fig. 5.3a. When the mass flow rate through the compressor is higher than the design value, the axial component of velocity is higher. With the blade angles and the blade speed remaining the same, the relative velocity vector at the rotor inlet, C_2 is no longer tangential to the blade surface at the inlet. This leads

---

[2] It must be kept in mind that the pressure rise follows a geometric progression. Hence, the pressure rise across 15 stages will be $\frac{1.1^{15} - 1}{1.1 - 1} \approx 32$.

[3] Centrifugal compressors are used in turboprop engines such as the PW100 from Pratt and Whitney even today. The interested reader may visit their web site www.pwc.ca for more details and pictures.

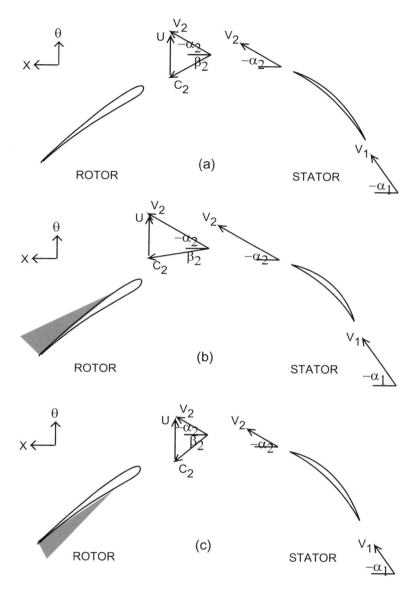

Fig. 5.3 Operation of a compressor stage at **a** design **b** greater than design and **c** less than design mass flow rate. Shaded areas indicate separated flow regions

to flow separation from the upper surface of the blade, which is shown in Fig. 5.3b as a shaded region. When the mass flow rate is less than the design value, the axial velocity component is also less and the relative velocity vector C_2 is once again not tangential to the blade surface at the inlet. In this case, the flow separates on the lower side of the blade surface, which is shown shaded in Fig. 5.3c. Flow separation on the blade surface leads to a condition known as *stalling*, which is detrimental to the performance of the compressor. It is thus desirable to operate within a narrow range of mass flow rates not too far above or below the design value.

During actual operation, slight variations in the inlet conditions are unavoidable. Since the pressure rises across the compressor in geometric progression, these variations are also magnified in the same manner. Consequently, while the low pressure stages may be operating near the design point, the last (high pressure) stages will be operating far away from the design point and hence will be stalled.

This difficulty is even more acute during start-up. At the time of start-up, the air inside the compressor has the same ambient density throughout. This value of the density is considerably lower than the value for which the high pressure stages are designed. Hence, the axial velocity at the later stages can become quite high, even reaching sonic velocity. Once this happens, these stages become *choked*, and the mass flow rate that they can handle is limited to the choking value. However, as the compressor continues to spin up, more mass is drawn in and the choked stages cannot handle the mass flow rate that is being passed from the front stages. In other words, the mass "piles up" in the later stages. This results in the so-called start-up problem. One way of getting around this piling up is to by-pass (bleed) some of the air coming from the upstream stages, so that the remaining air can go through the compressor. As the compressor warms up, the amount of air that is by-passed can be slowly decreased thereby allowing the compressor to come up to idling speed.

A closer inspection of the velocity triangles in Fig. 5.3 reveals that the departure of β_2 from the design value under off-design operating conditions is due to the fact that the flow angle α_2 and the blade speed U are held constant. In Fig. 5.3b, if α_2 were decreased appropriately to keep β_2 at the design value, stalling can be avoided entirely. Similarly, for the situation shown in Fig. 5.3c, α_2 can be increased appropriately to keep β_2 at the design value. The corresponding velocity diagrams for these cases are shown in Fig. 5.4. This strategy requires that the stator vanes be *movable*, so that α_2 can be varied. Of course, in actual operation, the range over which α_2 can be varied is restricted by other considerations. Nevertheless, this is a very effective strategy for controlling compressor stalling and is used commonly in aircraft engines.

The alternative strategy of varying the blade speed does not seem practical (or feasible) at first sight. For, the blade speed U at a point on the rotor equals $r\omega$ and the only means of varying U is to vary the angular velocity ω, which is not possible since the stages are all mounted on a *single* shaft. However, such a strategy would be possible if *multiple* shafts were used. This is explored in detail in the next chapter.

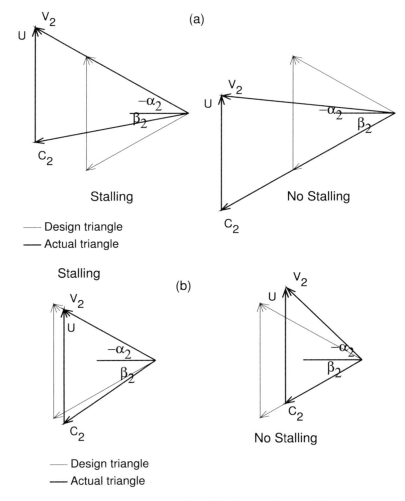

Fig. 5.4 Operation of a compressor stage under off-design operating conditions with moving guide vane. Mass flow rate **a** greater than design and **b** less than design mass flow rate. The design velocity triangle is shown in gray

5.1.3 Tip Clearance Control

One of the most challenging tasks during operation of a gas turbine engine is maintaining the clearance between the tip of the blades and the outer casing. This clearance should be as small as possible, to minimize leakage and to improve efficiency. Tip clearance control can be achieved through passive and active strategies.

In the first strategy, in some designs, the blade tips are manufactured with an abradable insert. As the blade expands during operation, the tips are ground off, providing a snug fit. An alternative approach is to make the seals (which are attached to the casing) abradable and make the blade tip abrasive. In this case, the blade tips will grind and machine the seal to the required clearance.

Active strategy essentially involves cooling (or, conversely, allowing to expand) the case and/or the disks on which the blades are mounted. Cool air from the low pressure stages is used to cool certain areas of the compressor case. This shrinks the case and reduces the blade tip clearance. The amount of cooling air used and the duration of cooling is adjusted actively during flight based on parameters such as altitude, rotor speed and rotor temperature. In some designs, the cooling air supply to the interior of the rotor (which is usually hollow) is controlled. A reduction in the cooling air supply results in an increase in the rotor temperature, which in turn allows the rotor to expand and fit snugly against the case.

5.1.4 Materials

The pressure and temperature range in the compressor section of the engine is from 1 to 40 atm and 200 to 1000 K. The compressor blades and the disk[4] in the high pressure section are thus exposed to high pressures and moderately high temperatures. In addition, the centrifugal stress, which is tensile in nature, is also high owing to the high rotational speed (typical rotational speeds are 8000–9000 rpm). This is more so in the low pressure section since the blade height is more here.

Thus, the requirements on the compressor blade and disk material are that it should have high tensile strength and also be able to resist fatigue.[5] The compressor disks and blades are usually made of titanium alloys on the cold side and heat resistant nickel-based alloys on the hotter end.

5.2 Combustor

Liquid fuel (usually JP4 or Jet A) is injected, mixed and burnt with air in the combustor. The combustor design should be able to achieve stable combustion over the entire range of idle, cruise and take-off throttle setting. Uniform temperature profile (both circumferential and radial) at the combustor exit is also another important requirement. Other issues such as emissions and lifetime of components are also important and these are discussed later.

---

[4] The rotor blades can be mounted in different ways. In one design, all the blades are mounted on a common rotor called a drum rotor which is driven by the shaft. In another design, the rotor blades are mounted on individual disks and the disks are bolted together and driven by the shaft.

[5] Fatigue is a weakening of the blade and disk material due to repeated cyclic loading. The cyclic loading arises from temperature variations as the engine is throttled through idle, take-off, climb, cruise, descent, reverse thrust and then idle again during each flight schedule.

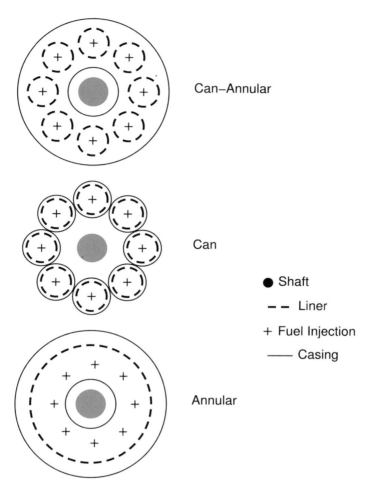

Fig. 5.5 Cross-sectional view of different types of combustors. Adapted from Mechanics and Thermodynamics of Propulsion by Hill and Peterson

Three types of combustor design have been used so far in gas turbine engines. These are the annular, can and can–annular combustors illustrated in Fig. 5.5. All three types have a casing, liner and fuel injectors. The liner has holes along its circumference and length to allow air to enter in a radially inward direction and mix with the combustion products. Some of the air also enters alongside the fuel in an axial direction.

In an annular combustor, there is an inner casing also and the combustion is confined between the liner and the inner casing. In the can combustor design, air intake, fuel injection and combustion take place in separate combustors cans. It is easy to see from Fig. 5.5 that each combustor can is a smaller version of the annular

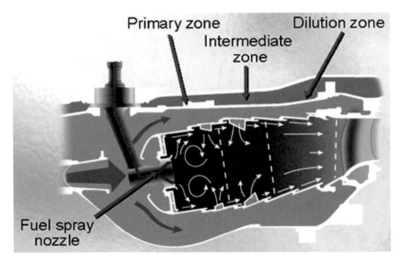

Fig. 5.6 Longitudinal section of a combustor can. *Courtesy* Rolls-Royce plc

combustor but without the inner casing. The can–annular design is a combination of the first two designs. The only difference is that the individual cans lack the outer casing and have only the liner and fuel injector(s). They are arranged along the circumference of a circle in between a common outer and inner casing and are connected to each other by a connecting tube called the flame tube (not shown in Fig. 5.5). Thus, the cans are not separated from each other by a casing as is the case in the can combustor design. Fuel injection in each can is accomplished individually but air intake is common for all the cans. The flame tube allows the flame to travel from one can to another at the time of starting. The flame is initiated using energy released from igniter plugs.

The can combustor allows the fuel–air ratio in the individual cans to be controlled well. In addition, owing to the modular design, any failure requires the replacement in a single can alone, not the entire combustor. However, synchronizing the ignition in the cans is difficult and consequently the non-uniformity in the outlet temperature distribution can be severe. This problem is mitigated to a large extent in the other two designs. However, failure of the liner, for instance, requires disassembling the entire combustor resulting in longer down times. Notwithstanding this, the can–annular combustor design is now used extensively in aircraft gas turbine engines, primarily because the pressure drop is the lowest in this design.

A longitudinal section of a combustor can is shown in Fig. 5.6. As shown in this figure, three zones, namely, primary (combustion) zone, intermediate (cooling) zone and dilution zone can be identified in a combustor can. The overall air–fuel ratio is a gas turbine engine can be as high as 60 : 1 or 70 : 1. Of the total amount of air that enters the combustor, about 20% enters the primary combustion zone in the axial direction alongside the fuel. Thus, the mixture in the core of the combustion zone is near stoichiometric (or slightly rich). The remaining 80% of the air is diverted radially outward around the liner. About 20% of the air enters the combustion zone, 40% enters the cooling zone and 20% enters the dilution zone through the holes in the liner. The diameter of the holes in the liner should be small enough to create the required pressure drop to generate sufficient turbulence and enhanced mixing of the air with the combustion products. Both the circumferential and axial location of the holes have a strong influence on the uniformity of the temperature profile at the combustor exit, since they control the extent of mixing and hence the peak temperature. The peak temperature at the combustor exit is about 1800 K in modern day combustors (Fig. 5.2). This value should be compared with the value obtained in Exercise problem 4 in Chap. 4. Turbulence and mixing are essential for good combustion, dilution of the hot products and for achieving a uniform temperature profile at combustor exit. However, the irreversibilities associated with turbulence and mixing can result in loss of stagnation pressure, which is undesirable. Therefore, a good combustor design is the outcome of an exercise in the optimization of these conflicting requirements.

One additional issue that arises with combustors in aircraft engines is the reduction in the rate of the chemical reactions with pressure (and hence altitude). Under design operating conditions, combustion reactions will be complete within combustion zone itself. At higher altitudes combustion will take a longer time to complete and hence will continue into the intermediate zone and perhaps even further downstream. Therefore, the combustor should be sufficiently long to accommodate such variations.

There are other practical issues with gas turbine combustors, such as fuel injection, atomization, mixing of fuel and air, flame holding and combustion instability that have not been discussed here. These are well beyond the scope of this book, and the reader should consult the books suggested at the end for discussions of these topics.

5.2.1 Materials

It can be seen from Fig. 5.2 that the combustor and the high pressure turbine stages operate at the highest pressure and temperature. Although the pressure is high in the combustor, there is no work interaction and it is stationary, i.e., non-rotating. Hence, the mechanical stress in the casing and the liner is mainly due to thermal loading. This loading, in addition to a mean component, can also have a time varying component arising out of the different phases during a flight schedule (mentioned earlier in connection with the compressor). From a metallurgical point of view, oxidation of the liner material is a serious issue owing to its proximity to the high temperature gases. Therefore, the primary requirements of the material for the combustor and liner are that it should be able to withstand high stresses due to severe thermal loading and resist oxidation. The liner and casing are manufactured from nickel-based alloys. Ceramic coated material and ceramic composites currently under development are likely to be used in the future.

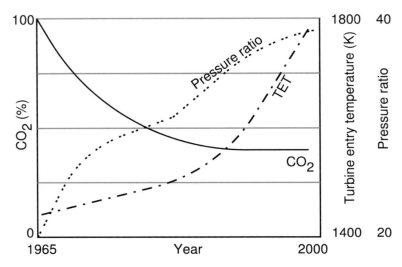

Fig. 5.7 Trends in turbine inlet temperature, pressure ratio and CO_2 emission. Data courtesy Rolls-Royce plc

Notwithstanding the superior quality of the materials used for its manufacture, the combustor liner needs additional protection, since the temperature of the hot gases (≈ 1800 K) are well above the limiting temperature of the underlying metal (≈ 1200 K). This is achieved through film cooling, wherein a thin layer of relatively cool air is maintained between the hot gases and the liner material. This avoids direct exposure of the liner material to the hot combustion gases.

5.2.2 Emissions

It will be shown in Chap. 7 that the efficiency of a turbojet engine and the thrust generated are dependent directly on the maximum pressure and temperature that can be realized in the engine. Consequently, a great amount of research and development work has gone into the ways and means of increasing the peak pressure and temperature.

Trends in the turbine inlet temperature and pressure ratio in aircraft engines over the years are shown in Fig. 5.7. Increase in maximum pressure has been achieved through improved compressor aerodynamics and increase in turbine entry temperature has been made possible through the use of better materials and better cooling methods for the combustor and the high pressure turbine stages.

While higher turbine entry temperature is desirable (and necessary) from a performance perspective, it has a detrimental effect from the emissions perspective. With emission norms becoming very stringent these days, it is a challenging task to

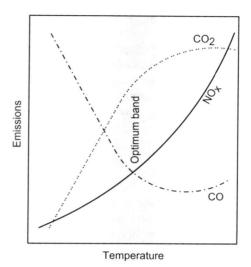

Fig. 5.8 Schematic variation of CO_2, CO and NO_x emission with temperature

satisfy the conflicting requirements of higher turbine entry temperature and lower emissions.

Combustion of a hydrocarbon fuel will, inevitably, release CO_2 as one of the major products. Since it is a greenhouse gas, the amount of CO_2 released should be minimal. This is especially important in the case of aircraft engines, since the CO_2 is being released at high altitudes. Improvements in combustion efficiency (and hence fuel economy) and thrust produced per kg of fuel, over the years has led to significant reductions in the amount of CO_2 released, as shown in Fig. 5.7. As mentioned earlier, these improvements can be attributed to increase in pressure ratio and turbine entry temperature. Also, combustion of the fuel tends to be more complete with increasing temperature and hence the amount of unburnt hydrocarbons present in the exhaust gases will be less.

Figure 5.8 shows that with increasing temperature, the amount of CO decreases initially owing to combustion being complete. However, at higher temperatures, the CO_2 formed during the combustion reaction dissociates into CO and O_2 according to the reaction $CO_2 \rightleftharpoons CO + \frac{1}{2}O_2$, and so the amount of CO in the exhaust begins to increase. In addition, with increasing temperature, the N_2 in the product gas mixture begins to react with the O_2 through reactions such as $\frac{1}{2}N_2 + \frac{1}{2}O_2 \rightleftharpoons NO$ to form oxides of nitrogen, namely, NO_x. Both the NO_x and the CO in the exhaust gases are very harmful to the environment. As shown in Fig. 5.8 there is a band of operating temperature, at which the emissions from the engine, considering the major pollutants, CO_2, CO and NO_x can be said to be optimal.

While the combustor can be optimized for a particular operating condition, say, cruise condition, optimizing for other operating conditions is a challenging task. In a typical flight schedule, the power requirements vary from minimum power during idling to maximum power at take-off. One concept that is being explored currently to address this design challenge is *staged combustion*. The staged combustor is a modular design with several small combustors operating in tandem, instead of a monolithic combustor. Each stage is operated under optimal conditions and as the power requirements of the engine change during flight, additional stages are either brought on line or turned off. Emissions are thus kept at an optimal level regardless of the power setting of the engine since the individual stages of the combustor are always operating at the optimal operating condition.

5.3 Turbine

The turbine produces the work required to drive the compressor. Since the flow expands as it goes through the turbine, the pressure gradient is favorable. Hence, the blade loading coefficient can be high for the turbine stages. A single turbine stage can produce enough work to drive several compressor stages and consequently the number of turbine stages is generally far fewer than the number of compressor stages. This is illustrated in Fig. 5.1. The rapid temperature and pressure drop is illustrated in Fig. 5.2. It should be noted that just enough work to drive the compressor is extracted from the fluid. The remaining enthalpy is used in the nozzle to generate thrust.

Tip clearance issues for turbines are the same as those of the compressor. Hence, similar strategies are used for control of tip clearance in turbines as well. Earlier turbine designs featured a shroud at the blade tip. The shroud is a circular metal band fixed to the rotor blades at their tips and served to prevent leakage past the tips as well as increase the structural rigidity of the blade assembly. Although it is very effective, the additional mass is a distinct disadvantage. With better materials and active tip clearance controls, shrouded rotors are not used today.

It is clear from Fig. 5.2 that the turbine stages (especially the HP stages) are subjected to high pressures and temperatures. In addition, unlike the combustor, the turbine rotor blades are also rotating at high speeds. In fact, the centrifugal load on a HP turbine blade can be several tons. As already mentioned, the temperature of the gases at the combustor exit (or turbine entry) is generally above the melting point of the turbine blade material. Hence, extreme care is required in the design, fabrication and operation of these stages. A two pronged strategy is generally used, wherein

- the blades are cooled as much as possible, and
- the blades are fabricated with better material to resist creep, fatigue and corrosion.

These are discussed next.

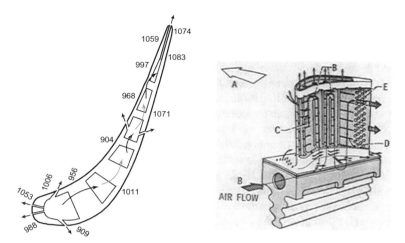

Fig. 5.9 Cut section of a turbine blade (left) showing channels for film cooling air and internal cooling passages. Representative values of temperatures (in °C) at different locations are also shown. Perspective view of the cooled turbine blade (right). *Source* www.aere.iastate.edu

5.3.1 Blade Cooling

Air from the later stages of the compressor is drawn for cooling the high pressure stator, rotor blades and disks.[6] It should be noted that this coolant air itself is at a temperature of 1000 K or so! In general, about 5–8% of the air from the compressor is used for cooling purposes and the coolant air does not participate in thrust generation. It is thus important to realize that cooling the turbine blades results in better efficiency and performance by allowing the turbine entry temperature to be higher but there is a thrust penalty. Ultimately, the amount of air used for cooling has to be weighed against this penalty.[7] The intermediate and low pressure turbine stages do not require any cooling as the temperature of the gases decreases considerably after passing through the high pressure stages, as is evident from Fig. 5.2.

The air drawn for cooling purposes is used in several different ways to cool the turbine blades. Some of the widely used methods are impingement cooling, film cooling and forced convection cooling. The cooled turbine blades are hollow with the coolant channels inside as shown in Fig. 5.9. Air from the compressor enters the coolant passage nearest the leading edge from the root and travels toward the tip. There are two reasons for this. Firstly, the heat transfer from the hot gas to the blade metal is the highest in this part of the blade cross-section owing to its proximity

---

[6] The stator blades, although not subject to centrifugal loads, require cooling for the same reasons as the combustor liner does—high thermal stresses and corrosion.

[7] It might seem that the amount of air required can be reduced by using cooler air from the low pressure stages rather than the hotter air from the high pressure stages. This is not practical, since high pressure is required to force the air through the intricate and small cooling passages.

to the stagnation point. Secondly, the temperature of the coolant air is the lowest as it has just entered the coolant passage and hence the amount of heat that can be transferred to it is the highest. The coolant air then travels from the tip to the root in the next passage and then in the opposite direction in the next passage (indicated by the arrows) and finally leaves through the trailing edge of the blade. This multi-pass (serpentine) convection cooling is quite effective in cooling the blade and maintaining the blade metal at acceptable temperature levels.

Small channels connecting the internal passage to the blade surface near the leading edge are also shown in Fig. 5.9. The coolant air at high pressure flows through these channels and onto the blade surface and maintains a thin film along the length of the blade. This insulates the blade metal surface from the hot gases. Additional coolant air is injected at a few more locations along the length of the blade to maintain the film. Representative temperature values at different locations on the blade surface (taken from Halila et al., NASA CR 167955, 1982) are also shown in Fig. 5.9. It can be seen that the temperatures are the highest near the leading and the trailing edge. The temperature decreases from the leading toward the middle and then begins to increase, owing to the decreasing blade thickness (and amount of blade metal). It emerges from Fig. 5.9 that the blade trailing edge region is the most vulnerable part of the blade section.

5.3.2   Materials

The turbine blade material should have high rotational strength and in the case of the high pressure stages, the blade material must be able to withstand high pressure and temperature. In addition, the following metallurgical considerations are important in the selection of appropriate materials for the turbine blades.

- Creep: Stretching of blades due to prolonged application of moderate loads at elevated temperatures
- Fatigue: Weakening due to repeated cyclic loading. The cyclic loading is due to temperature variations as the engine is throttled through idle, take-off, climb, cruise, descent, reverse thrust and then idle again during each flight schedule
- Corrosion: Mainly oxidation of the turbine metal (usually, nickel-base alloys with chromium and cobalt).

Normally, a failure (crack) is initiated along those grain boundaries of the blade metal which are oriented normal to the stress axis (which is along the radial direction). Hence, the rotor blades of the high pressure turbine are manufactured by a process known as the Directional Solidification Process (first introduced by Pratt and Whitney) which yields crystals aligned along the radial direction. The grain boundaries are thus parallel to the stress axis and so the blades last longer. Single crystals are even better as they do not have any grain boundaries at all.

Commonly used materials are nickel-based alloy for the disk and single crystal nickel-based alloy for the blades. Current trend is to have a thermal barrier coating for the blades of the high pressure stages. The coating is of a ceramic material and provides additional protection to the blade surface, which allows the turbine entry temperature to be increased further.

5.4 Nozzle

The primary function of the nozzle is to convert the enthalpy of the gases leaving the turbine into kinetic energy. Since this involves expansion of the flow, the pressure gradient is favorable. Also, the temperatures and pressures are not very high. For these reasons, the nozzle is relatively easy to design. On turbojet engines used for subsonic flights, the nozzle is a convergent one with a fixed throat cross-sectional area. If the engine has an afterburner (discussed below), then the nozzle has adjustable flaps (or segments) which allow the throat area to be changed when the afterburner is in operation.[8] If the flight Mach number is supersonic, then the nozzle has adjustable elements which allow it become a convergent divergent nozzle from a convergent one.

5.4.1 *Thrust Reversing*

Reverse thrust is generally used immediately after landing to help decelerate the aircraft. Although the main landing gear wheels have brakes, use of reverse thrust reduces the rolling distance after landing by almost a factor of 3.[9] Mechanisms to generate reverse thrust usually form a part of the nozzle subsystem assembly. Clam shell type thrust reverser is most widely used with turbojet engines. This consists of two buckets (clam shell shaped). In the stowed mode, they lie alongside and on opposite sides of the nozzle and in the deployed mode they come out and form a V-shaped surface which diverts the jet backwards (Fig. 5.10). This generates the reverse thrust necessary to decelerate the aircraft. Figure 5.10 also shows such a thrust reverser in the deployed mode on an actual aircraft with a turbojet engine.

5.5 Afterburning Turbojet Engine

Afterburners are used in turbojet engines for short duration thrust augmentation during takeoff from aircraft carriers, combat manoeuvres and for accelerating though the sound barrier. They are used almost exclusively in military aircraft. The Concorde

---

[8] See Worked Example 6.12 in Fundamentals of Gas Dynamics by V. Babu.

[9] Other devices such as parabrake (parachute aided braking) and tow wire braking are used in military aircrafts. Turboprops reverse the pitch of the propeller blades to achieve reverse thrust.

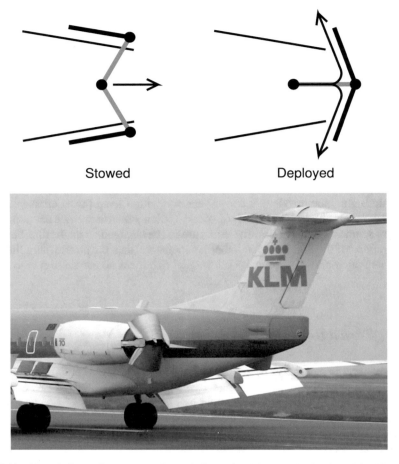

Fig. 5.10 Clam shell type thrust reverser in a turbojet engine. Photograph taken by Adrian Pingstone in May 2005 and released to the public domain

powered by four Rolls-Royce Olympus engines was the only civilian aircraft to have an afterburner (for the last reason). A schematic view of an afterburning turbojet engine compared with a regular turbojet engine is given in Fig. 5.11. It can be seen that the differences between the two are downstream of the tail cone in the nozzle. The afterburner has a long tail pipe terminating in an adjustable nozzle. Since the overall air–fuel ratio in a gas turbine is quite high, as already mentioned, considerable amount of oxygen is present in the exhaust gases. The basic idea behind the afterburner is to inject and burn additional fuel in this exhaust stream to generate additional thrust. The increase in the stagnation temperature of the exhaust gases allows more enthalpy drop in the nozzle and hence more thrust.

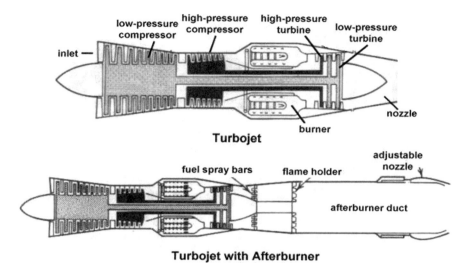

Fig. 5.11 Schematic of an afterburning turbojet engine. *Source* http://www.aerospaceweb.org/ question/propulsion/q0175.shtml

Fig. 5.12 Afterburner nozzle with adjustable flaps in the stowed position. *Source* http://www. solarnavigator.net

Fig. 5.13 Afterburner nozzle operating with increased throat area. Note that the flaps in the nozzle are fully deployed. *Courtesy* Rolls-Royce plc

It can be surmised from Fig. 5.11, the velocity of the exhaust gases at the location where the fuel is injected will be very high. Consequently, mechanical flame holders, called V-gutters are required to stabilize the combustion. These V-shaped flame holders create low speed recirculation regions behind them. These contain hot, slow moving product gases which tend to anchor the flame. The flame holders are located at a considerable distance downstream of the fuel injectors, to give enough time for the fuel to penetrate and mix with the high speed main flow. The tail pipe is also made long to allow combustion of as much fuel as possible inside the afterburner. Despite this, when the afterburner is in operation, the flame usually extends several nozzle diameters outside the tailpipe. The combustion efficiency of an afterburner is consequently not very high. However, the amount of fuel required to reach a given supersonic flight speed through a short duration afterburn can actually be less than that required to overcome the drag during the long climb without afterburning.

In a gas turbine engine, the Mach number at entry to the combustor is low (<0.3). Hence, compressibility effects are not significant. The amount of heat that can be added is limited only by metallurgical considerations. On the other hand, the flow velocity and the Mach number in the afterburner tail pipe are such that compressibility effects are significant. The amount of heat that can be added is limited now by "thermal choking".[10] In such a case, when the afterburner is in operation, the nozzle throat area has to be increased to get around this difficulty. This is usually accomplished through interlocking flaps (Figs. 5.12 and 5.13) which slide out around the regular nozzle when the afterburner is lit. This, of course, increases the mechanical complexity of the nozzle.

---

[10] See Chap. 4 in Fundamentals of Gas Dynamics by V. Babu.

Chapter 6
The Turbofan Engine

In the last chapter, the turbojet engine was discussed in detail. Historically, commercial aviation began toward the end of World War II with propeller-driven planes. It soon became apparent that the flight speeds of the propeller planes were limited to low subsonic values. The use of the gas turbine engine in combination with a propulsion nozzle, i.e., the turbojet engine, allowed higher flight speeds to be realized. Although still used for military applications, turbojets lack the efficiency required for commercial applications. This led to the evolution of the turbofan engine. These issues are explored in detail in this chapter.

6.1 Concept of Bypass Engine

The variation of efficiency with flight speed for different engines is shown in Fig. 6.1. The efficiency of the propeller, although very good at low flight speeds, falls off rapidly at high speeds. This is because the propeller speed increases with flight speed and the tip speed reaches supersonic values very quickly owing to the large propeller diameter. Once this happens, normal shock waves form at the tips. This leads to shock-induced flow separation due to the adverse pressure gradient from the pressure rise across the shock wave, which, in turn, reduces the efficiency drastically and also increases the noise.

With the birth of the turbojet engine, the limitation on the flight speed was removed entirely. Turbojets are capable of operating at high subsonic and supersonic flight speeds. However, their efficiencies are low. Also, the maximum pressure and temperature encountered in the turbojet engine are quite high, which brings with it many additional complexities, as discussed in the previous chapter. The exhaust from the turbojet engine contains combustion products (pollutants) that are harmful to the environment. Furthermore, the mixing of the high-speed, hot exhaust jet from the turbojet engine with the ambient air produces very high levels of noise. It should

Fig. 6.1 Schematic of the variation of efficiency for different engines. The shaded area represents the operating range for commercial aircraft. Adapted from Mechanics and Thermodynamics of Propulsion by Hill and Peterson

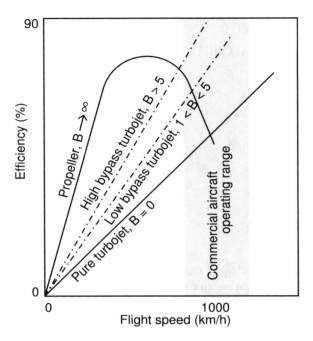

be noted that the noise from a propeller is mainly of mechanical origin whereas the noise from a turbojet is of fluid dynamic origin. Increasing fuel cost and increasing environmental awareness, viz., stringent norms for emissions and noise levels, motivated the development of a better alternative, namely, the bypass turbojet or, more popularly, the turbofan engine.

Propellers, in contrast to turbojets, do not suffer from the disadvantages mentioned above. Their efficiency is higher than that of turbojets except at high subsonic flight Mach numbers and their peak efficiency is quite high. The velocity change across a propeller is modest. There is no change in the temperature of the working fluid since there is no heat addition.[1] Consequently, the change in velocity of the fluid is not high and hence noise arising from high-speed jet mixing is absent. The turbofan engine has emerged as a better alternative to turbojets, mainly because it has retained all of these advantages of the propellers.

Accordingly, a significant portion of the overall thrust (as much as 85%) in a turbofan engine is generated in a manner similar to that in a propeller, i.e., by imparting a small velocity change to a large mass of air by means of a *fan*. The remainder of the thrust is generated in a conventional turbojet engine, referred to as the *core engine*. Thus, only a fraction of the total air that enters the turbofan engine goes through the core turbojet engine. The remainder is bypassed around the core and goes through the fan. The bypass ratio, which is defined as the ratio of the mass flow rate of bypass

---

[1] This does not mean that there are no emissions. The powerplant that drives the propeller will, of course, have its own emissions.

air to the mass flow rate of air going through the core engine, is a key performance parameter for the turbofan engine. The pure turbojet discussed in the previous chapter has a bypass ratio equal to zero, while the propeller has a value approaching infinity. Most commercial aircraft engines have bypass ratios around 5 or less (depending on the thrust rating). As shown in Fig. 6.1, the bypass turbojet has superior efficiency compared to the turbojet at all flight Mach numbers. More importantly, at high sub-sonic flight Mach numbers which are of interest for commercial aviation operation,[2] it has efficiency comparable to the peak efficiency of the propeller. Today, the high bypass turbofan engine is more efficient, quieter and environmentally friendly than any of the other alternatives and boasts of a fuel economy per passenger kilometer which rivals that of the most efficient automobiles. It has thus become the engine of choice for commercial aviation. There were unique design challenges that had to be overcome before this became a reality. These are discussed next.

6.2   Transonic Fan Stage

Although the fan in a turbofan engine is conceptually similar to a propeller, the similarity ends there. In design and deployment, the fan is substantially different from the latter. For instance,

- The fan is shrouded since it is placed inside the fan duct.
- The fan compresses the air through a modest pressure ratio. Thrust is generated by expanding this flow in a fan nozzle. In contrast, the propeller (which is not shrouded) generates thrust by imparting a velocity change to the flow, with the pressure remaining at the same value as the ambient (except in the vicinity of the propeller surface itself).

In a turbofan engine, the Mach number based on the absolute velocity of the flow approaching the fan is subsonic. The Mach number based on the relative velocity of the flow is also subsonic near the root and increases toward the tip. This is due to the fact that the blade speed and hence the relative velocity (it should be recalled from Chap. 3 that the relative velocity, $C = V - U$, where V is the absolute velocity and $U = r\omega$ is the blade speed) increase from root to tip. This, coupled with the fact that the speed of sound is the lowest at the fan inlet owing to the static temperature being the lowest, results in the Mach number being high here. Therefore, if the tip radius is high enough, there is a possibility that the relative Mach number at the blade tip can become supersonic.

The tip radius is determined by the bypass ratio (or, equivalently, the mass flow rate of air taken in by the engine). For a given value of thrust and altitude, the air mass flow rate required by a high bypass turbofan engine can be as much as two and

---

[2] Preferred flight Mach number for commercial operation is about 0.9–0.95. The corresponding altitude and flight speed are 12 kms and 1000 km/h. These will change depending upon the prevailing wind speed and direction.

a half to three times that of a turbojet engine.[3] Consequently, the frontal area *has* to be large enough to capture the required amount of air.[4] This leads to the inevitable conclusion that the fan tip radius will be high enough to cause the relative Mach number at the fan inlet to be supersonic over a certain part of the blade height near the tip.

The pressure rise across the fan stage is determined by the fan nozzle thrust requirement.[5] From Eq. 5.2, the pressure rise across the fan is given as

$$\frac{dP}{\rho} \approx -d\left(\frac{C^2}{2}\right) = -CdC,$$

where we have assumed the streamlines to be horizontal. This shows that the required pressure rise can be accomplished with a small change in relative velocity across the fan, dC, but with a large value for C itself. The latter requirement implies that, once again, the relative Mach number at the fan inlet is likely to be supersonic over part of the blade near the tip. Also, with the absolute velocity at inlet fixed, high values for C can be realized only with high values for U, or, equivalently, ω.

For a given rotational speed, as the fan tip radius increases, or *vice versa*, the centrifugal stress on the fan blade increases. Though the mass flow rate and pressure rise considerations demand high tip radius and rotational speed, material considerations demand exactly the opposite. Hence, the design of the turbofan engine is an optimization (or, compromise!) between these two competing factors.

6.2.1 Wide-Chord Fan Blade

As mentioned earlier, the absolute velocity of the flow at the fan inlet is the same across the height of the blade. Since the blade velocity increases from root to tip, the blade angle at the inlet changes correspondingly as illustrated in Fig. 6.2. The variation in the angle is quite high for a fan blade (and a propeller blade) owing to its height and so the blade is highly twisted. However, one important difference between the propeller and the fan is that the blades function in isolation in the former. In a fan, the pressure rise is accomplished by diffusion of the flow in the passage between the blades, as demonstrated earlier. Although the design procedure for the blade passage in a compressor is quite well established, the supersonic relative Mach number at the inlet near the tip makes the design a challenging task. For the fan tip diameters required in high bypass turbofan engines, the value for this Mach number can be as high as 1.6.

Diffusion of the subsonic flow approaching the fan for part of the blade height from the root is usually achieved in a *diverging* blade passage, with a double circular

---

[3] The airflow rate for the Rolls-Royce Trent engine during take-off is nearly 1000 kg/s.
[4] Modern high bypass turbofan engines have fan tip diameters approaching 2 m.
[5] Typical value of fan pressure ratio for high bypass ratio commercial engines is about 1.7.

Fig. 6.2 Velocity triangles at the fan inlet

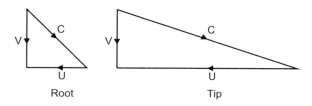

Root   Tip

arc (DCA) blade profile (Fig. 6.3). On the other hand, diffusion of the supersonic flow approaching the tip section to a subsonic Mach number requires a combination of external compression with oblique shocks, followed by internal compression in a *converging diverging* passage. As illustrated in Fig. 6.3, DCA blades are not suitable, since the supersonic flow will accelerate over the convex suction surface near the leading edge and trigger a normal shock at the mouth of the blade passage. Since this normal shock occurs at a value of relative Mach number which is higher than the value at the inlet, the loss of stagnation pressure is quite high and this decreases the aerodynamic efficiency of the fan considerably. Hence, a custom-designed blade profile is required in the tip section. The design and successful deployment of the custom-designed wide-chord fan blade was a key technology breakthrough in the development of the high bypass turbofan engine.

The custom-designed profiles of the fan blades in any production engine are proprietary and so they are not available in the open literature. Figure 6.4 shows a custom-designed profile that accomplishes the same task. The important features in the blade shape are the *concave* portion on the suction surface near the leading edge and the long chord.

Salient features of the flow field are shown at the bottom in Fig. 6.4. In contrast to the flow field shown in the previous figure for a DCA blade, there is now a bow shock ahead of the leading edge. This is due to the fact that the leading edge for the blade profile shown in Fig. 6.4 is rounded rather than being sharp as in Fig. 6.3. Although the Mach number decreases across the bow shock, it should be noted that the Mach number near the blade suction surface still increases to a value of 1.76 as it goes around the rounded leading edge. The concave portion on the suction surface beyond the leading edge decelerates the supersonic flow from a Mach number of 1.76 to a minimum value of 1.53 through a series of weak oblique shocks as shown in Fig. 6.4. Beyond the concave portion, as the blade profile becomes convex, expansion waves are generated and the flow accelerates to a maximum Mach number of 1.68. Just beyond this location, the flow encounters a passage shock, which is the lower part of the bow shock in front of the neighboring blade. The passage shock is shown highly simplified in this illustration. In reality, it is a complex shock structure, usually terminating in a normal shock on the suction surface. Consequently, there will be separated flow region(s) on the suction surface. Under steady inlet conditions, the location and extent of these regions as well as the location of the passage shock can be controlled. The mixed supersonic–subsonic flow is gradually diffused in the

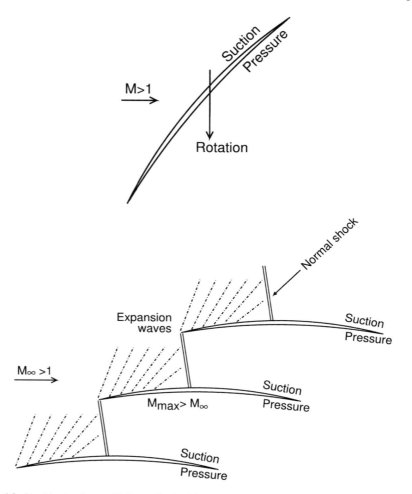

Fig. 6.3 Double circular arc blade profile (top) for subsonic compression and the salient features of the flow field for supersonic inlet relative Mach number (bottom). Note the diverging blade passage

remainder of the blade passage to the required subsonic value at the exit. This gradual diffusion necessitates a long blade passage and hence a long chord for the blade profile.

6.2.2   Multi-spool Engine Technology

If we recall from the previous chapter that rotational speeds of the order of 10,000 rpm are necessary for the efficient functioning of the HP compressor stages, it becomes clear that the fan stage *cannot* be mounted on the same shaft as the rest of the compres-

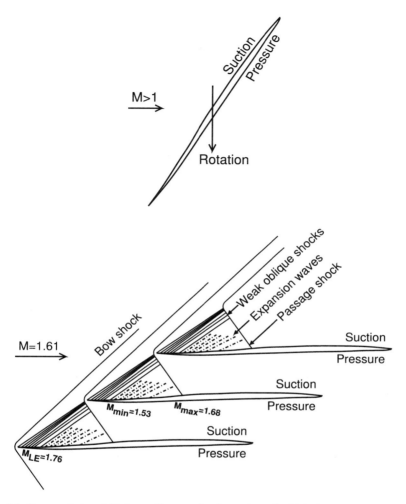

Fig. 6.4 Custom-designed blade profile near the tip section of a high through flow axial flow compressor rotor (top) and the salient features of the flow field (bottom). Adapted from Tweedt et al., Experimental investigation of the performance of a supersonic compressor cascade, NASA TM 100879, 1986

sor stages, since the centrifugal stress on the fan blade would be well above the yield stress of most materials. This realization led to another technological breakthrough, namely, the multi-spool or multi-shaft engine. The idea here is to have multiple hollow shafts one inside the other, instead of a single solid shaft. This allows each shaft to run at a different rpm. Turbofan engines today have two or three spools. The three-spool engine was pioneered by Rolls-Royce and introduced into commercial service in the RB211 engine family. An exploded view of the three spools in a three-spool engine is given in Fig. 6.5 (see cover illustration also).

Fig. 6.5 Exploded view of a three-spool turbofan engine. *Courtesy* Rolls-Royce plc

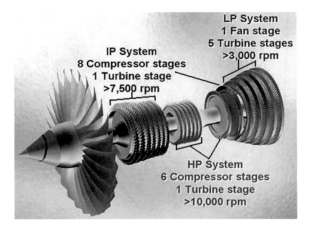

The compressor/turbine combination has three modules, namely, the LP, IP (intermediate pressure) and the HP module. The LP module consists of the single fan stage driven by 5 turbine stages and spins at 3000 rpm. The IP module consists of 8 compressor stages driven by a single turbine stage and spins at 7500 rpm. The HP module consists of 6 compressor stages driven by a single turbine stage and spins at 10,000 rpm. It can be seen from Fig. 6.5 that the HP module shaft has the largest diameter with the IP and LP module shafts having progressively smaller diameters.

One of the advantages of the multi-spool design is that the individual modules can operate at their optimum level under design as well as off-design operating conditions since each module is spinning at a different (optimal) speed. The usefulness of this strategy was alluded to earlier in Chap. 5 and is illustrated in Fig. 6.6. It can be seen that stalling of the stages can be controlled effectively by changing the rotational speed. The more the number of spools, the more robust this control will be. However, increasing the number of spools increases the mechanical complexity of the engine. Rolls-Royce engines feature the maximum number of spools, namely, three. Two-spool engines in use today have movable guide vanes in addition to achieve the same flexibility and control. Multi-spool engines also tend to be lighter, shorter in length and more rigid.

6.2.3   Fan Blade Material

As discussed above, the multi-spool concept allows the fan to run at a substantially lower speed than the rest of the compressor stages, which reduces the centrifugal stress on the blade.[6] This is desirable from a structural perspective. The wide-chord

---

[6] Notwithstanding this, the centrifugal load on a fan blade during take-off (maximum power) can be as high as 100 tons.

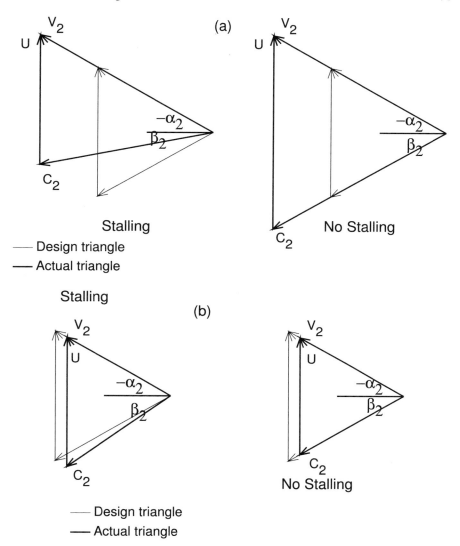

Fig. 6.6 Operation of a compressor stage under off-design operating conditions with multiple spools. Mass flow rate **a** greater than design and **b** less than design mass flow rate. The design velocity triangle is shown in gray

Fig. 6.7 Fan in a
Rolls-Royce Trent 1000
engine. *Courtesy*
Rolls-Royce plc

blade shape, on the other hand, requires the blade to be thin in cross-section and consequently, tall and slender. This is not desirable from a structural perspective, since tall, slender blades are prone to failure, unless they are sufficiently rigid. Such conflicting requirements led to yet another technological breakthrough, namely, the titanium fan blade.

Rolls-Royce pioneered the hollow, wide-chord fan blade made from titanium panels. The hollow core is filled with a honeycomb matrix of titanium to give the blade strength and rigidity. These blades are used for their high-end RB211 family and Trent family of engines. In the smaller engines, the fan blades are solid titanium. These blades, although tall and slender, are exceptionally strong. In the older turbofan engines, part-span shrouds, also called "snubbers", were used in fan blades to give additional structural rigidity. These snubbers, however, increase aerodynamic losses in the blade passage. Owing to continuous improvements in fabrication methods, modern fan blades can be manufactured with the required strength without snubbers.

The outcome of all these technological breakthroughs, namely, the transonic fan, in a Rolls-Royce Trent engine is shown in Fig. 6.7. All the features discussed above,

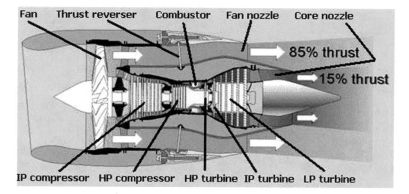

Fig. 6.8 Illustration of a three-spool turbofan engine. *Courtesy* Rolls-Royce plc

namely, the twisted blade passage for the diffusion of transonic flow, large tip diameters for capturing the required mass flow rate and the long, slender blades can be discerned from this figure.

An important safety issue that arises in the case of a turbofan engine is foreign object damage (particularly bird hit). Since the frontal area of the engine is large, the probability of such an event happening is also higher. The fan is especially prone owing to its large diameter. Any damage to a blade can have an adverse effect on the performance of the fan and any mechanical imbalance resulting from the damage can potentially have serious consequences. Hence, the blade material, in addition to being able to withstand the centrifugal stresses, must also be able to withstand damage due to foreign object impact. In the extreme event that a blade is dislodged from the hub, a special fan blade containment strategy utilizing a circumferential band made of kevlar or titanium in the shroud ensures that the fan blade is contained within the engine.

6.3 Operation of a Turbofan Engine

A cut-away view of a three-spool turbofan engine is given in Fig. 6.8 (see cover illustration also). The single-stage fan also serves as the LP compressor. The air that goes through the core region of the fan (approximately, one-third of the blade height) is directed to the core turbojet engine. This air then goes through the IP compressor and the HP compressor. The combustor cans and the various turbines are also shown in this figure. The number of stages in each turbine/compressor configuration and its speed is the same as in Fig. 6.5. Since the fan generates much of the thrust, the power required to drive the fan is also more and so as many as six LP turbine stages are utilized for this purpose. The core (hot) nozzle generates about 15% of the total thrust, which is typical of high bypass turbofan engines.

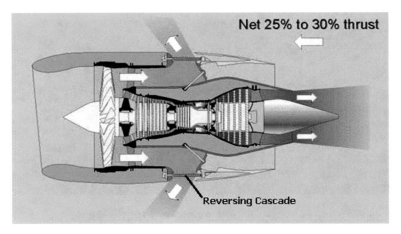

Fig. 6.9 Illustration of thrust reversing in a turbofan engine. *Courtesy* Rolls-Royce plc

The bypass air undergoes compression in the fan, with pressure ratio being generally of the order of 1.6 or 1.7. There is also a set of exit guide vanes in the fan duct to straighten and diffuse the flow some more. The bypass air is then expanded in the fan (cold) nozzle, which generates as much as 85% of the overall thrust. In some engines, part of the cold fan air is mixed with the hot gases from the core engine, before expansion in the core nozzle. This has the effect of reducing the jet noise from the high speed jet issuing from the core nozzle.

6.3.1   Thrust Reversing

Since the fan generates a large part of the thrust, thrust reversing mechanism in a turbofan engine is deployed in the fan stream, not in the core nozzle. One such mechanism is shown in the deployed position in Fig. 6.9 (the mechanism is shown in Fig. 6.8 in the stowed position). When the thrust reverser is deployed, a door on the fan shroud slides out, exposing a cascade of airfoils along the circumference of the shroud. Simultaneously, a door inside the fan duct blocks the fan air from going to the fan nozzle and forcing it instead to exit through the cascade. The fan air is redirected in the forward direction by the cascade, which generates the reverse thrust required to decelerate the aircraft quickly.

Chapter 7
Thrust Calculations

In the last two chapters, basic concepts and theory behind the design and operation of turbojet and turbofan engines were discussed. In this chapter, the thermodynamic operation of these engines is discussed. Expressions for the thrust developed as well as the thrust specific fuel consumption are derived. A procedure for calculating these quantities for a given set of operating conditions is also outlined.

7.1 Derivation of Expression for Thrust and TSFC

7.1.1 Turbojet Engine

Consider an aircraft flying at a speed V_a at a given altitude where the static pressure and temperature are P_a and T_a, respectively. Figure 7.1 illustrates this for the engine in a frame of reference in which the aircraft is stationary and the air approaches with velocity V_a. Air enters the engine with a velocity V_a and the exhaust gases leave the engine with a velocity V_e. Fuel enters the engine with a mass flow rate equal to $\dot{m}_f$. Considering the engine as a control volume, momentum balance for the control volume can be written as

$$\left(P_a + \rho_a V_a^2\right) A_a + \int_{a,\text{surface}}^{e} P\, dA = \left(P_e + \rho_e V_e^2\right) A_e. \tag{7.1}$$

The integral term in the momentum equation arises due to the pressure force on the wall. For the geometry shown in Fig. 7.1, this is the force exerted on the engine[1]

---

[1] It should be remembered that thrust force acts in the negative x-direction and drag force in the positive x-direction. Hence, a negative value for this integral would imply thrust and a positive value, drag.

© The Author(s), under exclusive license to Springer Nature Switzerland AG 2022
V. Babu, *Fundamentals of Propulsion*,
https://doi.org/10.1007/978-3-030-79945-8_7

Fig. 7.1 Schematic
illustration of a turbojet
engine

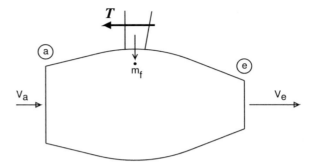

and hence the airframe. To facilitate the evaluation of this force, we define a quantity
called Impulse Function, I at any x-location as follows:

$$I = (P + \rho V^2)A. \tag{7.2}$$

The net force exerted is the difference between the value of the impulse function at
the exit and the inlet, viz.,

$$T = I_e - I_a.$$

It is easy to see from this equation and Eq. 7.1 that $T = \int_{a,\text{surface}}^{e} P dA$. From the
above equation, the thrust produced by the engine is

$$T = \left(PA + \rho V^2 A\right)_e - \left(PA + \rho V^2 A\right)_a.$$

This can be written as

$$T = (PA + \dot{m}V)_e - (PA + \dot{m}V)_a$$

where $\dot{m} = \rho V A$ is the mass flow rate of air. Strictly speaking, the mass flow rate at
the exit will be greater than that at the inlet by $\dot{m}_f$. However, as mentioned earlier in
Chap. 5, the overall air–fuel ratio on a mass basis for aircraft gas turbine engines is
about 60:1 or higher. Hence, $\dot{m}_f$ is much smaller than $\dot{m}$ and has been neglected in
the above equation. If the static pressures in this equation are measured relative to
the ambient pressure, then the thrust is given by

$$T = \dot{m}(V_e - V_a) + (P_e - P_a)A_e. \tag{7.3}$$

   The first term in this equation is called *momentum thrust* and $\dot{m}V_a$ term with a
negative sign is called *intake momentum drag*.[2] The second term in the above equation
is called *pressure thrust* and is non-zero when the pressure of the fluid at the exit
is not equal to the ambient pressure. In this case, the fluid is "under-expanded" and

---

[2] In the case of a rocket engine, air is not taken in through any inlet and so this term will be absent.

it expands further outside the nozzle and equilibrates with the ambient conditions a few nozzle diameters downstream of the exit. The expansion is accomplished across an expansion fan centered at the nozzle lip. The jet swells initially as it comes out of the nozzle and expands, but shrinks afterwards due entrainment of the ambient air and equalization of static pressure.

When the fluid expands outside the nozzle, the thrust generated is less than optimum. Although it can be seen from Eq. 7.3 that the pressure thrust is non-zero in this case, this gain is more than offset by the reduction in momentum thrust due to the reduced velocity of the fluid at the exit. If in a particular propulsion nozzle this loss becomes too high, then the solution is to replace it with a convergent divergent nozzle. This allows expansion of the fluid beyond the sonic state inside the nozzle and hence recovers the lost thrust. In propulsion applications, convergent nozzles can be used without severe penalty on the thrust as long as $P_{0,\text{nozzle}}/P_a < 3$. Beyond this value, convergent divergent nozzles have to be used to utilize the momentum thrust fully.

It is easy to see from Eq. 7.3 that for a given V_e, as the flight speed V_a is increased, thrust decreases. Also, a decrease in the density of the incoming air with everything else fixed, results in a decrease in the thrust, since the air mass flow rate into the engine, $\dot{m} = \rho_a V_a A_a$ decreases. Conversely, an increase in the density at the inlet increases the thrust owing to increased mass flow rate through the engine. Since $P = \rho RT$, it follows that a decrease in inlet temperature alone or an increase in the pressure alone, increases the thrust.[3] Thus, aircraft engines usually develop more thrust on a cold day than on a hot day. Conversely, a decrease in pressure alone or an increase in temperature alone, decreases the thrust. With increasing altitude, both ambient temperature and pressure decrease, with pressure decreasing faster than temperature. The overall effect is a gradual reduction in thrust with increasing altitude, assuming that V_e and V_a remain the same and $P_e = P_a$.[4] Beyond 10 kms, temperature remains constant and so the reduction in thrust is drastic and hence this is the preferred altitude for commercial subsonic flights.

Thrust specific fuel consumption (TSFC) is defined as the mass flow rate of fuel required to generate a given thrust. Thus

$$\text{TSFC} = \frac{\dot{m}_f}{\mathcal{T}} = \frac{\dot{m}_f/\dot{m}}{\mathcal{T}/\dot{m}}. \tag{7.4}$$

The quantity in the numerator in the second equality is the reciprocal of the air–fuel ratio, and the quantity in the denominator is called the specific thrust. In practical calculations, the specific thrust is of more significance and interest than the thrust itself, as it ultimately determines the engine cross-sectional area. Since specific thrust

---

[3] This effect was exploited in earlier days by spraying water or water-alcohol mixture into the inlet. Evaporation of the liquid reduces the air temperature augmenting the thrust during take-off.

[4] In reality, the thrust required to maintain level flight at a given altitude under the prevailing wind conditions is achieved by adjusting the fuel flow rate and the air mass flow rate (by changing the rpm of the HP turbine/compressor system) to match the throttle setting. This is explored later in Worked Example 7.1.

is the thrust generated per unit mass flow rate of air captured by the engine, the higher the specific thrust, the lesser the air mass flow rate required to generate a given thrust. This translates directly to lesser engine cross-sectional area.

From Eq. 7.3, we can write

$$\frac{T}{\dot{m}} = V_e - V_a + (P_e - P_a)\frac{A_e}{\dot{m}}. \tag{7.5}$$

Since $\dot{m} = \rho_e V_e A_e$ and $\rho_e = P_e / R_g T_e$, the above equation can be simplified as

$$\frac{T}{\dot{m}} = V_e - V_a + (P_e - P_a)\frac{R_g T_e}{P_e V_e}. \tag{7.6}$$

Here, R_g is the particular gas constant for the product gas mixture at the combustor exit. In addition to the specific thrust and TSFC, overall efficiency is also important. This is defined as

$$\eta_o = \frac{T V_a}{\dot{m}_f H_{\text{fuel}}} = \frac{(T/\dot{m}) V_a}{(\dot{m}_f/\dot{m}) H_{\text{fuel}}} \tag{7.7}$$

where H_{fuel} is the calorific value of the fuel.[5]

At first sight, it appears from Eq. 7.6 that three quantities, namely, P_e, T_e and V_e have to be known in order to calculate the specific thrust. In reality, however, only two quantities, namely, the stagnation pressure and temperature at the nozzle inlet have to be known, since P_e can be evaluated by making use of the fact that the propulsion nozzle is a convergent nozzle in aircraft engines. The nozzle, in this case, can operate in only one of two modes—unchoked or choked, depending upon whether the ratio of the stagnation pressure at the nozzle inlet to the ambient static pressure is less than the critical value or not.[6] Accordingly, in the first case, $P_e = P_a$. In the second case, the Mach number at exit is unity and this allows P_e to be determined. The procedure for calculating the specific thrust and the TSFC for an ideal turbojet engine is outlined later in this chapter.

7.1.2 Turbofan Engine

In the case of a turbofan engine, thrust is generated from the hot (core engine) and cold (fan) stream. By evaluating the impulse function at the fan nozzle exit, hot

---

[5] Substituting for $T/\dot{m}$ from Eq. 7.3 and assuming that $P_e = P_a$, we get

$$\eta_o = \frac{(V_e - V_a) V_a}{(\dot{m}_f/\dot{m}) H_{\text{fuel}}}.$$

As the flight speed V_a is varied keeping the jet velocity V_e fixed, it can be easily shown from this expression that η_o exhibits a maximum when $V_a = V_e/2$.

[6] See Chap. 6 in Fundamentals of Gas Dynamics by V. Babu.

nozzle exit and the inlet, the thrust developed can be written as

$$T = \dot{m}_H(V_{e,H} - V_a) + (P_{e,H} - P_a)A_{e,H}$$
$$+ \dot{m}_C(V_{e,C} - V_a) + (P_{e,C} - P_a)A_{e,C}.$$

Here, subscripts C and H denote cold and hot streams, respectively. The specific thrust is then equal to

$$\frac{T}{\dot{m}} = \frac{\dot{m}_H}{\dot{m}}(V_{e,H} - V_a) + \frac{A_{e,H}}{\dot{m}}(P_{e,H} - P_a)$$
$$+ \frac{\dot{m}_C}{\dot{m}}(V_{e,C} - V_a) + \frac{A_{e,C}}{\dot{m}}(P_{e,C} - P_a)$$

where the total mass flow rate into the engine $\dot{m} = \dot{m}_C + \dot{m}_H$. Since the bypass ratio $B = \dot{m}_C/\dot{m}_H$, it is easy to show that $\dot{m}_C = \dot{m}B/(B+1)$ and $\dot{m}_H = \dot{m}/(B+1)$. The above expression can then be written as

$$\frac{T}{\dot{m}} = \frac{1}{B+1}(V_{e,H} - V_a) + \frac{A_{e,H}}{\dot{m}}(P_{e,H} - P_a)$$
$$+ \frac{B}{B+1}(V_{e,C} - V_a) + \frac{A_{e,C}}{\dot{m}}(P_{e,C} - P_a). \tag{7.8}$$

Since $\dot{m}_C = \rho_{e,C}V_{e,C}A_{e,C}$ and $\rho_{e,C} = P_{e,C}/RT_{e,C}$, we can write

$$\frac{A_{e,C}}{\dot{m}} = \frac{B}{B+1}\frac{RT_{e,C}}{P_{e,C}V_{e,C}}.$$

Similarly

$$\frac{A_{e,H}}{\dot{m}} = \frac{1}{B+1}\frac{R_g T_{e,H}}{P_{e,H}V_{e,H}}$$

Thus, Eq. 7.8 can be written as

$$\frac{T}{\dot{m}} = \frac{1}{B+1}\left(V_{e,H} - V_a + \frac{R_g T_{e,H}}{P_{e,H}V_{e,H}}(P_{e,H} - P_a)\right)$$
$$+ \frac{B}{B+1}\left(V_{e,C} - V_a + \frac{RT_{e,C}}{P_{e,C}V_{e,C}}(P_{e,C} - P_a)\right). \tag{7.9}$$

The expression in the first bracket is the core engine thrust, and the expression in the second bracket is the fan thrust. The similarity in the expression for specific thrust between the turbofan engine, Eq. 7.9 and the turbojet engine, Eq. 7.6 is noteworthy. In the limit as $B \to 0$, the expression for a turbojet engine is recovered. The comments made earlier in connection with the exit pressure for a turbojet engine are applicable here as well. Thus, for the cold and the hot nozzle, when the ratio of the stagnation pressure at the nozzle inlet to the exit static pressure is less than the critical value,

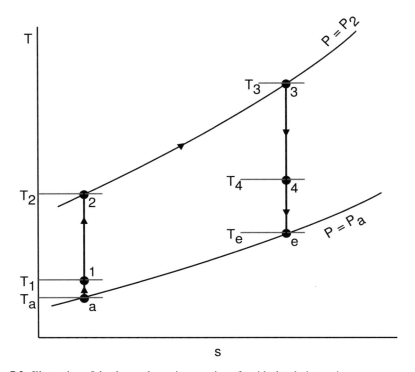

Fig. 7.2 Illustration of the thermodynamic operation of an ideal turbojet engine

then the flow is not choked and the exit pressure is equal to the ambient pressure; otherwise, the flow is choked.

The expression for thrust specific fuel consumption of the turbofan engine is the same as in Eq. 7.4 for the turbojet engine. However, unlike in a turbojet engine, $\dot{m}_f/\dot{m}$ is not the reciprocal of the air–fuel ratio in the case of the turbofan engine. The air–fuel ratio in this case is equal to $\dot{m}_H/\dot{m}_f = 1/(B+1)\,\dot{m}/\dot{m}_f$.

7.2   Thermodynamics of an Ideal Turbojet Engine

The thermodynamic processes undergone by the air that goes through an ideal turbojet engine are given in Fig. 7.2. In this figure, processes a–1 (intake), 1–2 (compressor), 3–4 (turbine) and 4–e (nozzle) are assumed to be isentropic. Also, heat addition in the combustor is assumed to be a constant pressure process (2–3) without any loss of stagnation pressure. In other words, all the processes have been assumed to take place without any irreversibilities. Hence, on the whole, we have assumed the engine to be ideal. Both the static state and the corresponding stagnation state are shown, since work and heat interactions have to be calculated using the latter.

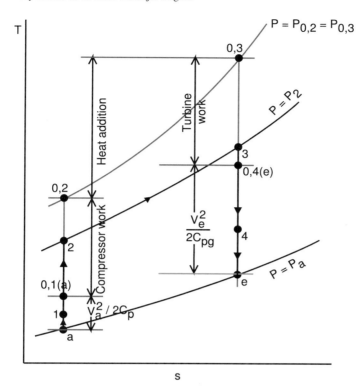

Fig. 7.3 Heat, work interactions and thrust in an ideal turbojet engine with exit pressure $P_e = P_a$

The variation of the static pressure and temperature of the fluid across the engine can be discerned by following the process line a–1–2–3–4–e. This should be correlated with the information given in Fig. 5.2. It should also be noted that the exit pressure P_e has been shown to be equal to P_a in Fig. 7.2.

If we apply the energy equation 2.17 to the compressor, combustor and the turbine, it is easy to see that the work or heat interaction is dependent only on the change in stagnation temperature. In the nozzle and the intake, the stagnation temperature remains constant with only conversion of static enthalpy to velocity or *vice versa* taking place. The heat and work interactions as well as the velocities V_e and V_a are shown in Fig. 7.3. It is important to note that the turbine work and the compressor work are equal.

The stagnation pressure and temperature increase across the compressor due to work addition, and decrease across the turbine due to (an equal amount of) work extraction. Stagnation temperature increases across the combustor due to heat addition, but without any loss of stagnation pressure. Stagnation quantities do not change across the inlet and the nozzle since there is no heat or work interaction. In the nozzle, the static enthalpy of the fluid is converted into kinetic energy and *vice versa* in the

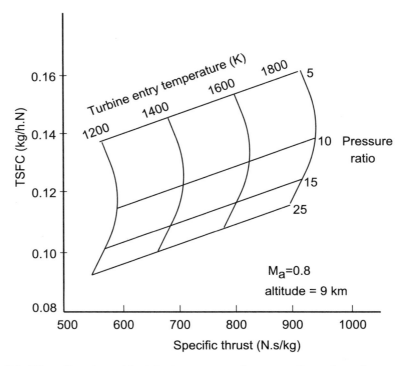

Fig. 7.4 Effect of varying turbine inlet temperature and pressure ratio on the performance of a turbojet engine. Reproduced from Gas Turbine Theory by Saravanamuttoo, Rogers and Cohen

intake. In the case of subsonic flight speeds, the pressure rise in the intake due to the deceleration of the fluid is not significant. However, for supersonic flight speeds the pressure rise can be substantial and so designing the intake to achieve this with minimal loss of stagnation pressure is crucial. This is discussed in detail in the next chapter. It can be seen from Fig. 7.2 that the working fluid does not execute a cyclic process, and hence the notion of thermal efficiency of a heat engine is not applicable to the turbojet engine.[7]

From a performance perspective, an engine should have high specific thrust and low TSFC. Optimization of the thermodynamics of the turbojet engine to achieve this goal requires varying the turbine inlet temperature (T_{03}) and the compressor pressure ratio (P_{02}/P_{01}), which are the only variable operational parameters. The outcome of such an exercise for a particular altitude and cruise Mach number is shown in Fig. 7.4.

Increasing the turbine entry temperature while keeping the pressure ratio fixed causes state point 3 to move to the right along the isobar $P = P_2$ (Fig. 7.3). Simultaneously, state points 4 and e also move to the right, the latter along the isobar

---

[7] Of course, an air standard efficiency can always be defined.

$P = P_e$. Heat addition and hence fuel flow rate are higher. Since the two isobars, $P = P_2$ and $P = P_e$ diverge from each other as one moves to the right (Fig. 7.3) and the turbine work remains the same, V_e is also higher. Therefore, increasing the turbine entry temperature alone increases the specific thrust. The effect on thrust specific fuel consumption cannot be accurately discerned from Fig. 7.3. However, in general, TSFC increases as shown in Fig. 7.4.

Increasing the pressure ratio while keeping the turbine entry temperature fixed causes state points 3, 4 and e to move to the left along their respective isobars. Since the static temperature at the compressor exit is higher now, heat addition and fuel flow rate are lower. On the other hand, compressor work and hence turbine work are both higher. This usually results in very marginal change in specific thrust. This, combined with the reduced fuel flow rate results in lower TSFC (Fig. 7.4), which is desirable.

While higher turbine entry temperature and pressure ratio may be desirable from thermodynamic considerations, it must be borne in mind that the discussion in Chap. 5 shows that material and metallurgical considerations dictate otherwise.

7.3   Thermodynamics of an Ideal Turbojet Engine with Afterburning

The thermodynamic processes undergone by the air that goes through the ideal turbojet engine with afterburning are given in Fig. 7.5. Processes a–1, 1–2, 2–3 and 3–4 are the same as those in a turbojet engine. When the afterburner is not in operation, the hot gases from the turbine will undergo expansion in the nozzle as usual (process 4–e, shown in gray). With the afterburner in operation, more heat is added to the hot gases in the afterburner as shown by the idealized constant pressure process 4–5. The stagnation temperature at the afterburner exit (state point 5), T_{05} can be as high as twice the turbine entry temperature. The gases then undergo expansion in the nozzle (process 5–e). When the afterburner is in operation, the nozzle is almost always choked and this is explicitly indicated in Fig. 7.5 with $P_e > P_a$. The nozzle exit area also has to be higher in this case due to the increase in stagnation temperature.

7.4   Component Efficiencies

In our discussion of the thermodynamic operation of the turbojet so far, we have assumed all the components to be ideal and without any irreversibilities. In an actual engine, there will always be irreversibilities and it is important to take into account their effect on the performance of the components. It is usually difficult to calculate this, since the irreversibilities in an actual engine arise from a combination of factors such as viscosity, dissipation, heat loss and chemical reactions. Hence, it is customary to define a suitable performance metric for each component and use a value for this metric which is less than that of the ideal case, while calculating thrust and TSFC.

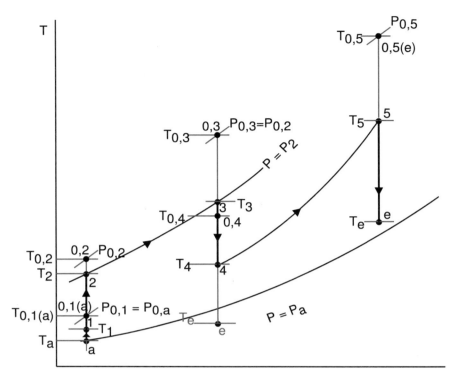

Fig. 7.5 Illustration of the thermodynamic operation of an ideal turbojet engine with afterburning

This is discussed next for each of the components in a turbojet engine. The same station numbering scheme as in Fig. 7.2 is used here.

7.4.1   Inlet

The actual diffusion process a–1 in the inlet in the presence of irreversibilities is shown in Fig. 7.6 as a dashed line. The static state and the corresponding stagnation states at the beginning and end of the diffusion process are also shown in this figure. As mentioned earlier, there is no heat or work interaction in the inlet, i.e., $T_{0,1} = T_{0,a}$ and so the effect of the irreversibilities is to increase the entropy, and hence a loss of stagnation pressure, i.e., $P_{0,1} < P_{0,a}$.

This suggests a performance metric for the inlet, namely, the stagnation pressure recovery, $P_{0,1}/P_{0,a}$. For an ideal inlet, stagnation pressure recovery is unity.[8] Alternatively, we can define an efficiency for the inlet as

---

[8] An equivalent metric, the stagnation pressure loss, is also frequently used. This is defined as $1 - \frac{P_{0,1}}{P_{0,a}}$.

Fig. 7.6 Illustration of the
thermodynamic operation of
the inlet with irreversibilities

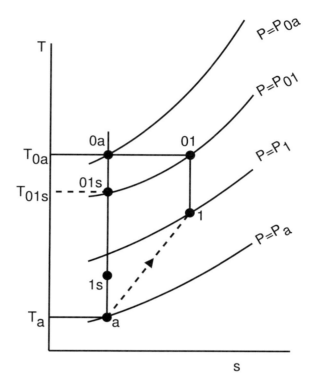

$$\eta_{inlet} = \frac{T_{0,1s} - T_a}{T_{0,1} - T_a}. \tag{7.10}$$

Here, $T_{0,1s}$ is the stagnation temperature corresponding to state $1s$ at the end of the
isentropic compression process a–$1s$, which is such that $P_{0,1s} = P_{0,1}$. For an ideal
inlet, $\eta_{inlet} = 1$.

It should be noted that both these metrics compare the thermodynamic performance of the actual process against an isentropic process. Since the primary function
of the inlet is to achieve a static pressure rise by decelerating the air that enters the
engine, it might seem that a performance metric involving the velocity of the air is
more desirable. For such a metric, the ideal case would be one where the flow is
decelerated to zero velocity, thereby converting *all* of the momentum of the fluid
into a pressure rise. This, of course, is a physically unrealistic situation. Hence, any
performance metric for the inlet should assess, **not** the pressure rise itself, but **how**
the pressure rise is achieved.

Fig. 7.7 Illustration of the thermodynamic operation of the compressor with irreversibilities

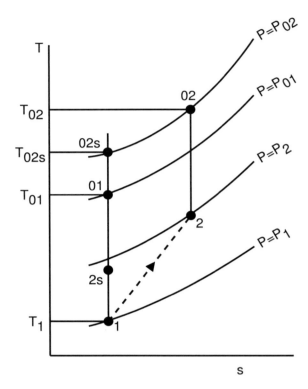

7.4.2 Compressor

The actual compression process 1–2 in the compressor is shown in Fig. 7.7 as a dashed line. The static state and the corresponding stagnation states at the beginning and end of the compression process are also shown in this figure. The compressor pressure ratio is defined as $P_{0,2}/P_{0,1}$. It can be seen from this illustration that the compressor is able to achieve the desired pressure ratio but does so in an irreversible manner. Any irreversibility usually manifests itself as a loss of stagnation pressure. In this case, it manifests itself in the form of increased work input for achieving a given stagnation pressure rise. An efficiency for the compressor can thus be defined as

$$\eta_{comp} = \frac{T_{0,2s} - T_{0,1}}{T_{0,2} - T_{0,1}}. \tag{7.11}$$

Here, $T_{0,2s}$ is the stagnation temperature corresponding to state $2s$ at the end of the isentropic compression process 1–2s, which is such that $P_{0,2s} = P_{0,2}$. For an ideal compressor, $\eta_{comp} = 1$.

7.4.3 Combustor

It is well known that addition of heat to a compressible flow, while increasing the stagnation temperature is always accompanied by a loss of stagnation pressure, even if the heat addition process is reversible.[9] Hence, the loss of stagnation pressure arises from the increase in entropy due to the heat addition, *not* due to any irreversibility. On the other hand, good mixing of the fuel and air is essential for good combustion, but this only contributes further to the loss of stagnation pressure since mixing is a highly irreversible process. Accordingly, the fractional loss of stagnation pressure in the combustor, $\Delta P_{0,\text{comb}} = (P_{0,2} - P_{0,3})/P_{0,2}$ is customarily used to quantify stagnation pressure loss in the combustor. In the ideal case, this would be equal to zero.

In addition, a combustion efficiency can also be defined for the combustor based on the notion that not all of the fuel injected undergoes complete combustion within the combustor. The challenges associated with this issue were discussed in Sect. 5.2. The combustion efficiency is defined as the ratio of the amount of fuel that undergoes complete combustion in the combustor to the amount of fuel injected. Hence

$$\eta_{\text{comb}} = \frac{\dot{m}_{f,\text{ideal}}}{\dot{m}_{f,\text{actual}}}. \tag{7.12}$$

7.4.4 Turbine

The actual expansion process 3–4 in the turbine is shown in Fig. 7.8 as a dashed line. The static state and the corresponding stagnation states at the beginning and end of the expansion process are also shown in this figure. The expansion in the turbine takes place across a pressure ratio of $P_{0,3}/P_{0,4}$. It can be seen from this illustration that, for a given stagnation pressure ratio, the turbine produces less work due to the presence of irreversibilities. Turbine efficiency can thus be defined as

$$\eta_{\text{turbine}} = \frac{T_{0,3} - T_{0,4}}{T_{0,3} - T_{0,4s}}. \tag{7.13}$$

Here, $T_{0,4s}$ is the stagnation temperature corresponding to state $4s$ at the end of the isentropic expansion process 3–4s, which is such that $P_{0,4s} = P_{0,4}$. For an ideal turbine, $\eta_{\text{turbine}} = 1$.

It is customary to define a mechanical efficiency, η_{mech} for the turbine to account for losses during the transmission of power from the turbine to the compressor. From Fig. 7.7, the actual work input to the compressor can be seen to be $\dot{m}C_p(T_{0,2} - T_{0,1})$

---

[9] Mach numbers at entry to the combustor in aircraft gas turbine engines are typically less than 0.3, since the the flow velocity is also quite low, as otherwise, a stable flame cannot be maintained. In addition, static temperature (see Fig. 5.2) and hence the speed of sound are both quite high. Therefore, compressibility effects are usually negligible.

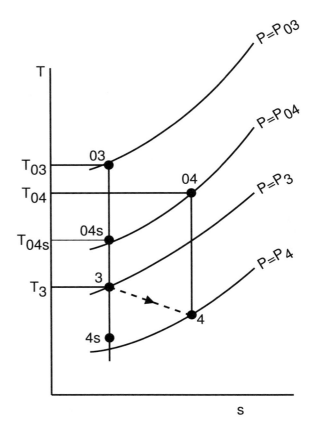

Fig. 7.8 Illustration of the thermodynamic operation of the turbine with irreversibilities

and from Fig. 7.8, the actual work output from the turbine is equal to $\dot{m}C_{pg}(T_{0,3} - T_{0,4})$. Mechanical efficiency of the turbine can thus be defined as

$$\eta_{mech} = \frac{\dot{m}C_p(T_{0,2} - T_{0,1})}{\dot{m}C_{pg}(T_{0,3} - T_{0,4})} = \frac{C_p(T_{0,2} - T_{0,1})}{C_{pg}(T_{0,3} - T_{0,4})}. \tag{7.14}$$

In the ideal case, $\eta_{mech} = 1$.

7.4.5   Nozzle

The actual expansion process 4–e in the nozzle is shown in Fig. 7.9 as a dashed line. The static state and the corresponding stagnation states at the beginning and end of the expansion process are also shown in this figure. As mentioned earlier, the function of the nozzle is to convert the static enthalpy of the fluid at state 4 into kinetic energy at the exit. Since there is no work or heat interaction, $T_{0,e} = T_{0,4}$. The

Fig. 7.9 Illustration of the thermodynamic operation of the nozzle with irreversibilities

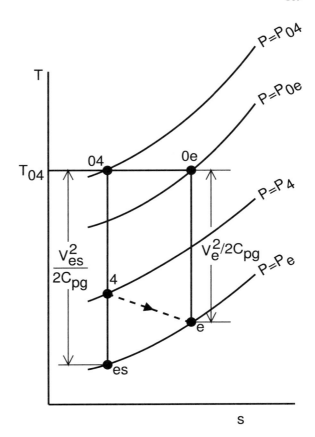

nozzle efficiency is given as

$$\eta_{\text{nozzle}} = \frac{T_{0,e} - T_e}{T_{0,e} - T_{es}} = \frac{V_e^2}{V_{es}^2}. \tag{7.15}$$

Here, state *es* at the end of the isentropic expansion process 4–*es* is such that the **static** pressure $P_{es} = P_e$. This ensures that differences in the exit velocity are due entirely to differences in the irreversibility in the expansion process and not due to differences in the exit static pressure. For an ideal nozzle, $\eta_{\text{nozzle}} = 1$.

All the efficiencies defined above (except for the combustor and the mechanical efficiency of the turbine) are the so-called isentropic efficiencies, since an appropriately defined isentropic process between the thermodynamic states of the fluid at the inlet and exit of the component has been taken as the ideal case. For turbomachinery such as the compressor and turbine, since the compression (or expansion) takes place across many stages and not a single stage, the notion of a polytropic efficiency is more appropriate. This takes into account the fact that the change in stagnation temperature across the machine is different when evaluated as the cumulative sum of the stagnation temperature change in each stage, than what is indicated in Figs. 7.7 and 7.8. Use of the polytropic efficiency instead of the isentropic efficiency will certainly affect the numerical values in the calculations somewhat, but we will use the latter definition throughout. The interested reader is referred to *Gas Turbine Theory* by H. I. H. Saravanamuttoo, G. F. C. Rogers and H. Cohen for further details.

7.5   Calculation of Specific Thrust and TSFC for a Turbojet Engine

In this section, a step-by-step procedure for calculating the specific thrust (Eq. 7.6) and TSFC (Eq. 7.4) for a turbojet engine is discussed. The data required to start the calculation is given below under different categories.

Flight Conditions V_a , T_a , P_a

Operating Conditions   Pressure ratio $PR = P_{0,2}/P_{0,1}$
                                                Turbine entry temperature $T_{0,3}$ (or $\dot{m}_f$)

Component efficiencies η_{inlet}, η_{comp}, $\eta_{turbine}$, η_{nozzle},
                                                η_{comb}, $\Delta P_{0,comb}$

Properties   Air (γ , C_p , R),
                                                Combustion gases (γ_g , C_{pg} , R_g),
                                                Heating value of fuel, H_{fuel}

7.5.1   Specific Thrust

As already discussed, to evaluate the specific thrust, it is enough if we determine $P_{0,4}$ and $T_{0,4}$. To this end, we start with $P_{0,a}$ and $T_{0,a}$ and then calculate P_0 and T_0 sequentially at state points 2, 3 and finally 4.

State a

From the given flight conditions, we can get

$$P_{0,a} = P_a \left(1 + \frac{\gamma - 1}{2} M_a^2\right)^{\frac{\gamma}{\gamma - 1}}$$

and

$$T_{0,a} = T_a \left(1 + \frac{\gamma - 1}{2} M_a^2\right)$$

where $M_a = V_a / \sqrt{\gamma R T_a}$ is the flight Mach number.

State 1

At the compressor inlet, $T_{0,1} = T_{0,a}$. From Eq. 7.10,

$$\frac{T_{0,1s}}{T_a} = 1 + \eta_{\text{inlet}} \left(\frac{T_{0,1}}{T_a} - 1 \right)$$

Since states a and $01s$ are connected by an isentropic process (Fig. 7.6), we can write

$$\frac{P_{0,1s}}{P_a} = \left(\frac{T_{0,1s}}{T_a} \right)^{\frac{\gamma}{\gamma - 1}}.$$

If we substitute for T_{01s} from above, we get

$$P_{0,1} = P_{0,1s} = P_a \left[1 + \eta_{\text{inlet}} \left(\frac{T_{0,1}}{T_a} - 1 \right) \right]^{\frac{\gamma}{\gamma - 1}}.$$

State 2

Since the compressor pressure ratio is known, $P_{0,2} = \text{PR} \times P_{0,1}$. From Eq. 7.11,

$$\frac{T_{0,2}}{T_{0,1}} = 1 + \frac{1}{\eta_{\text{comp}}} \left(\frac{T_{0,2s}}{T_{0,1}} - 1 \right).$$

Since states 01 and $02s$ lie on the same isentrope (Fig. 7.7), we can write

$$\frac{T_{0,2s}}{T_{0,1}} = \left(\frac{P_{0,2s}}{P_{0,1}} \right)^{\frac{\gamma - 1}{\gamma}}.$$

Therefore

$$T_{0,2} = T_{0,1} \left[1 + \frac{1}{\eta_{\text{comp}}} \left(\text{PR}^{\frac{\gamma - 1}{\gamma}} - 1 \right) \right].$$

Here, we have used the fact that $P_{0,2s} = P_{0,2}$.

State 3

Since the fractional stagnation pressure loss in the combustor is known, the stagnation pressure at the combustor exit $P_{0,3} = P_{0,2} (1 - \Delta P_{0,\text{comb}})$. Stagnation temperature at the combustor exit, $T_{0,3}$ is also known.

State 4

Since the turbine produces just enough work to run the compressor, application of the energy equation 2.17 to the turbine and the compressor gives,

$$\dot{m} C_p \left(T_{0,2} - T_{0,1} \right) = \eta_{\text{mech}} (\dot{m} + \dot{m}_f) C_{pg} \left(T_{0,3} - T_{0,4} \right).$$

If we use the fact that $\dot{m}_f \ll \dot{m}$, then this equation can be simplified to read

$$T_{0,4} = T_{0,3} - \frac{1}{\eta_{mech}} \frac{C_p}{C_{pg}} (T_{0,2} - T_{0,1}).$$

From Eq. 7.13, it can be shown that

$$\frac{T_{0,4s}}{T_{0,3}} = 1 - \frac{1}{\eta_{turbine}} \left(1 - \frac{T_{0,4}}{T_{0,3}}\right)$$

Since the states 04 and 04s are on the same isentrope (Fig. 7.8), it follows that

$$P_{0,4s} = P_{0,3} \left(\frac{T_{0,4s}}{T_{0,3}}\right)^{\frac{\gamma_g}{\gamma_g - 1}}.$$

If we substitute for T_{04s} from above, we get

$$P_{0,4} = P_{0,4s} = P_{0,3} \left[1 - \frac{1}{\eta_{turbine}} \left(1 - \frac{T_{0,4}}{T_{0,3}}\right)\right]^{\frac{\gamma_g}{\gamma_g - 1}}.$$

State e

As there is no heat or work interaction in the nozzle, $T_{0,e} = T_{0,4}$. If the expansion process 4–e in the nozzle is isentropic, then the critical value for the exit static pressure would simply be

$$P_{e,\text{crit}} = P_{0,4} \left(\frac{2}{\gamma_g + 1}\right)^{\frac{\gamma_g}{\gamma_g - 1}}. \tag{7.16}$$

Since the expansion process is not isentropic, an expression for the critical pressure has to be developed taking into account the nozzle efficiency.

When the exit static pressure is equal to the critical value, the exit Mach number is equal to unity and so

$$V_{e,\text{crit}}^2 = \gamma_g R_g T_{e,\text{crit}}.$$

Since $T_{e,\text{crit}} = \frac{2}{\gamma_g + 1} T_{0,e}$, we can write

$$V_{e,\text{crit}}^2 = \frac{2\gamma_g R_g}{\gamma_g + 1} T_{0,e}.$$

From Eq. 7.15, we can then calculate

$$V_{es,\text{crit}}^2 = \frac{1}{\eta_{nozzle}} \frac{2\gamma_g R_g}{\gamma_g + 1} T_{0,e}.$$

Hence

$$T_{es,\text{crit}} = T_{0,4} \left(1 - \frac{1}{2C_{pg}} \frac{1}{\eta_{\text{nozzle}}} \frac{2\gamma_g R_g}{\gamma_g + 1} \right)$$

where we have used the fact that $T_{0,e} = T_{0,4}$. Since $C_{pg} = \frac{\gamma_g R_g}{\gamma_g - 1}$, this expression simplifies to

$$T_{es,\text{crit}} = T_{0,4} \left(1 - \frac{1}{\eta_{\text{nozzle}}} \frac{\gamma_g - 1}{\gamma_g + 1} \right).$$

Since states es and 04 lie on the same isentrope (Fig. 7.9),

$$\frac{P_{0,4}}{P_{e,\text{crit}}} = \left(\frac{T_{0,4}}{T_{es,\text{crit}}} \right)^{\frac{\gamma_g}{\gamma_g - 1}}$$

where we have used the fact that $P_{es,\text{crit}} = P_{e,\text{crit}}$, as these states lie on the same isobar (Fig. 7.9). Substituting for $T_{es,\text{crit}}$ from above, we are finally led to

$$P_{e,\text{crit}} = P_{0,4} \left(1 - \frac{1}{\eta_{\text{nozzle}}} \frac{\gamma_g - 1}{\gamma_g + 1} \right)^{\frac{\gamma_g}{\gamma_g - 1}}. \tag{7.17}$$

This is the expression for the critical pressure ratio in a convergent nozzle in the presence of irreversibilities. It should be noted that Eq. 7.17 reduces to Eq. 7.16 when $\eta_{\text{nozzle}} = 1$, as it should.

If the ambient pressure $P_a > P_{e,\text{crit}}$, then $P_e = P_a$. Otherwise, $P_e = P_{e,\text{crit}}$. Once P_e is known, T_e can be evaluated as follows:

$$T_e = T_{0,e} - \frac{V_e^2}{2C_{pg}}$$

$$= T_{0,e} - \eta_{\text{nozzle}} \frac{V_{es}^2}{2C_{pg}} \quad \text{(from Eq. 7.15)}$$

$$= T_{0,e} - \eta_{\text{nozzle}} \left(T_{0,4} - T_{es} \right) \quad \left(\text{since } T_{0,4} = T_{es} + \frac{V_{es}^2}{2C_{pg}} \right)$$

$$= T_{0,4} \left[1 - \eta_{\text{nozzle}} \left(1 - \frac{T_{es}}{T_{0,4}} \right) \right]$$

$$= T_{0,4} \left[1 - \eta_{\text{nozzle}} \left(1 - \left(\frac{P_e}{P_{0,4}} \right)^{\frac{\gamma_g - 1}{\gamma_g}} \right) \right]. \tag{7.18}$$

In the last expression, we have taken $P_{es} = P_e$ and also used the fact that states es and 04 lie on the same isentrope (Fig. 7.9).

Once T_e is known, the exit velocity is given as

$$V_e = \sqrt{2C_{pg}(T_{0,e} - T_e)}. \tag{7.19}$$

The specific thrust can now be evaluated.

7.5.2 TSFC

To evaluate thrust specific fuel consumption, we start by applying the energy equation 2.17 to the combustor,

$$\dot{m} C_p T_{0,2} = (\dot{m} + \dot{m}_f) C_{pg} T_{0,3} - \dot{m}_f H_{\text{fuel}}.$$

Upon rearranging this, we get

$$\dot{m}_f = \frac{\dot{m} \left(C_{pg} T_{0,3} - C_p T_{0,2} \right)}{H_{\text{fuel}} - C_{pg} T_{0,3}}.$$

This expression for the fuel mass flow rate is for the ideal case and so it should be written as

$$\dot{m}_{f,\text{ideal}} = \frac{\dot{m} \left(C_{pg} T_{0,3} - C_p T_{0,2} \right)}{H_{\text{fuel}} - C_{pg} T_{0,3}}.$$

Hence, from Eq. 7.12,

$$\dot{m}_{f,\text{actual}} = \frac{\dot{m}_{f,\text{ideal}}}{\eta_{\text{comb}}}.$$

Thrust specific fuel consumption can now be calculated.

7.6 Thermodynamics of an Ideal Turbofan Engine

The thermodynamic operation of an ideal twin spool turbofan engine is shown in Fig. 7.10. As before, process a-1 represents diffusion in the inlet. Compression in the fan is represented by 1–2 and expansion in the fan (cold) nozzle by 2–eC. Processes 2–3–4–5–6–eH take place in the core engine. Since this is a twin spool engine, 4–5 represents expansion in the HP turbine which drives the core engine compressor and 5–6 represents expansion in the LP turbine which drives the fan. Process 6–eH represents expansion in the core engine (hot) nozzle. For the sake of illustration, the exit pressure in both the hot and cold nozzle is shown to be the same as the ambient pressure.

Stagnation states corresponding to the respective static states in Fig. 7.10 are shown in Fig. 7.11. As before, by applying the energy equation to each component, the work interaction, heat interaction or enthalpy conversion can be evaluated. These

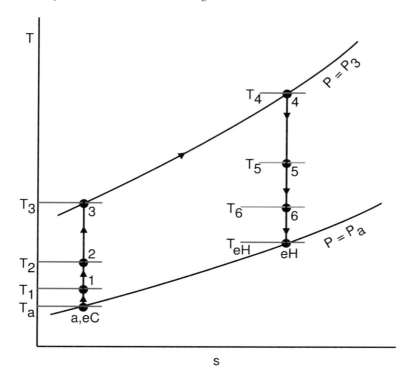

Fig. 7.10 Illustration of the thermodynamic operation of an ideal twin spool turbofan engine

are indicated in Fig. 7.11 for the core engine components. As shown in this figure, the HP turbine work is equal to the HP compressor work, since the HP turbine drives the HP compressor. The LP turbine work is equal to the fan work. However, the latter is not explicitly shown in Fig. 7.11. Although the work interactions for the LP turbine and the fan are equal, the change in the stagnation temperature is not the same between the two, since the mass flow rate and C_p are different for the fan (cold) stream and the core engine (hot) stream.

To begin with, let us assume that the bypass ratio, overall pressure ratio and the turbine inlet temperature are fixed. If the fan pressure ratio is increased now, it can be seen from Fig. 7.11 that the fan work (and hence the LP turbine work) increases and the HP compressor work (and hence the HP turbine work) decreases. Owing to

Based on our discussion so far, it is clear that there are four parameters that control the operation of the turbofan engine. These are the bypass ratio (B), fan pressure ratio ($P_{0,2}/P_{0,1}$), overall pressure ratio ($P_{0,3}/P_{0,1}$) and the turbine inlet temperature ($T_{0,4}$). It may be recalled that only the last two parameters are relevant for the turbojet engine. Optimization of the turbofan engine is thus more involved. A qualitative discussion based on Fig. 7.11 to illustrate some of the basic ideas is given next. For a detailed discussion, the reader should refer to the books given in the list of Suggested Reading.

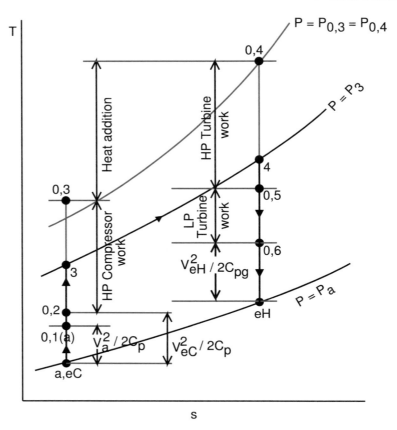

Fig. 7.11 Heat, work interactions and thrust in an ideal turbofan engine with exit pressure $P_e = P_a$

the difference in mass flow rates between the fan stream and core engine stream, the increase in the energy extracted in the LP turbine is more than the decrease in the HP turbine and so the hot thrust decreases. On the other hand, since the fan pressure ratio is higher, the fan thrust is also higher. The net effect due to the increase in the fan pressure ratio is an increase in the overall thrust. As the fan pressure ratio is increased further, the energy extracted from the LP turbine becomes higher and eventually the reduction in the hot nozzle exit velocity is so high that the reduction in the core engine thrust exceeds the increase in the fan thrust.[10] Thus, there exists an optimum value for the fan pressure ratio for a given set of values for the other parameters. Since the heat addition has been kept the same, the fuel flow rate also remains the same. Thus, TSFC also attains an optimum simultaneously. An inspection of Fig. 7.11 reveals

---

[10] As the fan pressure ratio is increased, $P_{0,3}$ increases and $P_{0,6}$ decreases. Since the ambient pressure is fixed, this will have the effect of choking the fan nozzle and increasing the fan thrust. Simultaneously, the hot nozzle becomes unchoked (if it were choked to begin with) and the momentum and pressure thrust from the hot nozzle both decrease.

that the optimum value for the fan pressure ratio will be higher for higher turbine inlet temperature. Optimum value for TSFC is also correspondingly higher.

On the other hand, if the bypass ratio is varied, keeping the other parameters fixed, it can be easily shown (by partially differentiating Eq. 7.8 with respect B) that the overall thrust decreases. Although the fan thrust increases, the reduction in the core engine thrust is more and so the overall thrust decreases. Increasing the bypass ratio diverts more air through the fan nozzle. Since the mass flow rate through the core engine is less, the decrease in stagnation quantities across the LP turbine has to be more, in order to develop the same amount of work. As a result, the reduction in the hot thrust is higher than before (with increasing fan pressure ratio). This can also be demonstrated as follows.

If we differentiate Eq. 7.8 with respect to B, keeping everything else fixed, we get

$$\frac{1}{\dot{m}}\frac{\partial T}{\partial B} = -\frac{1}{(B+1)^2}(V_{e,H} - V_a) + \left(\frac{1}{B+1} - \frac{B}{(B+1)^2}\right)(V_{e,C} - V_a).$$

This can be simplified as

$$\frac{1}{\dot{m}}\frac{\partial T}{\partial B} = -\frac{1}{(B+1)^2}(V_{e,H} - V_{e,C}).$$

Since $V_{eH} > V_{eC}$, there is a reduction in thrust with increasing bypass ratio.

The optimum values for these parameters obtained from thermodynamic considerations alone have to be evaluated further in the context of metallurgical and material considerations, as well as operational considerations (whether civil or military).

7.7 Component Efficiencies—Turbofan Engine

We have already discussed component efficiencies for different components in the turbojet engine in Sect. 7.3. Since these are directly applicable to the core engine in a turbofan, they need not be repeated here. Irreversibilities in the operation of the fan and fan nozzle can be handled in the same manner as that of the compressor and the hot nozzle. However, the expressions given before, namely, Eqs. 7.11, 7.13 and 7.15 must be used with caution since the station numbering for the turbofan engine (Fig. 7.10) is different from that of the turbojet engine.

7.8   Calculation of Thrust and TSFC for a Twin Spool Turbofan Engine

In this section, a step-by-step procedure for calculating the specific thrust (Eq. 7.9) and TSFC (Eq. 7.4) for a turbofan engine is discussed. The data required to start the calculation is given below under different categories.

Flight Conditions V_a, T_a, P_a

Operating Conditions   Overall pressure ratio $\mathsf{PR} = P_{0,3}/P_{0,1}$,
Fan pressure ratio $\mathsf{FPR} = P_{0,2}/P_{0,1}$,
HP turbine entry temperature $T_{0,4}$ (or $\dot{m}_f$)

Component efficiencies η_{inlet}, η_{comp}, $\eta_{turbine}$, η_{nozzle},
η_{comb}, $\Delta P_{0,comb}$,
η_{fan}

Properties   Air (γ, C_p, R),
Combustion gases (γ_g, C_{pg}, R_g),
Heating value of fuel, H_{fuel}

It should be noted that fan pressure ratio, fan efficiency η_{fan} are additionally specified. Although a cold (fan) nozzle efficiency can also be defined separately, we will assume that the isentropic efficiency of the cold and the hot nozzle to be the same, without any loss of generality.

7.8.1   Specific Thrust

As already discussed, to evaluate the specific thrust, it is enough if we determine $P_{0,2}$, $T_{0,2}$ and $P_{0,6}$, $T_{0,6}$. To this end, we follow the same strategy as before and start with $P_{0,a}$ and $T_{0,a}$ and then calculate P_0 and T_0 sequentially at state points 1, 2, 3, 4, 5 and finally 6.

Since conditions at entry to fan (state point 1) are the same as for the turbojet engine, we have $T_{0,1} = T_{0,a}$ and

$$P_{0,1} = P_a \left[1 + \eta_{inlet} \left(\frac{T_{0,1}}{T_a} - 1 \right) \right]^{\frac{\gamma}{\gamma-1}}.$$

State 2

Since the fan pressure ratio is known, $P_{0,2} = \mathsf{FPR} \times P_{0,1}$. From Eq. 7.11,

$$\frac{T_{0,2}}{T_{0,1}} = 1 + \frac{1}{\eta_{\text{fan}}} \left(\frac{T_{0,2s}}{T_{0,1}} - 1 \right).$$

Since states 01 and 02s lie on the same isentrope (Fig. 7.7), we can write

$$\frac{T_{0,2s}}{T_{0,1}} = \left(\frac{P_{0,2s}}{P_{0,1}} \right)^{\frac{\gamma-1}{\gamma}}.$$

Therefore

$$T_{0,2} = T_{0,1} \left[1 + \frac{1}{\eta_{\text{fan}}} \left(\text{FPR}^{\frac{\gamma-1}{\gamma}} - 1 \right) \right].$$

Here, we have used the fact that $P_{0,2s} = P_{0,2}$.

State 3

Since the overall pressure ratio is known, $P_{0,3} = PR\ P_{0,1} = (\text{PR}/\text{FPR})\ P_{0,2}$. From Eq. 7.11,

$$\frac{T_{0,3}}{T_{0,2}} = 1 + \frac{1}{\eta_{\text{comp}}} \left(\frac{T_{0,3s}}{T_{0,2}} - 1 \right).$$

Since states 02 and 03s lie on the same isentrope, we can write

$$\frac{T_{0,3s}}{T_{0,2}} = \left(\frac{P_{0,3s}}{P_{0,2}} \right)^{\frac{\gamma-1}{\gamma}}.$$

Therefore

$$T_{0,3} = T_{0,2} \left[1 + \frac{1}{\eta_{\text{comp}}} \left(\left\{ \frac{\text{PR}}{\text{FPR}} \right\}^{\frac{\gamma-1}{\gamma}} - 1 \right) \right].$$

Here, we have used the fact that $P_{0,3s} = P_{0,3}$.

State 4

Since the fractional stagnation pressure loss in the combustor is known, the stagnation pressure at the combustor exit $P_{0,4} = P_{0,3}\left(1 - \Delta P_{0,\text{comb}}\right)$. Stagnation temperature at the combustor exit $T_{0,4}$ is also known.

State 5

Since the HP turbine produces just enough work to run the compressor, application of the energy equation 2.17 to the HP turbine and the compressor gives

$$\dot{m}_H C_p \left(T_{0,3} - T_{0,2} \right) = \eta_{\text{mech}} (\dot{m}_H + \dot{m}_f) C_{pg} \left(T_{0,4} - T_{0,5} \right).$$

If we use the fact that $\dot{m}_f \ll \dot{m}_H$, then this equation can be simplified to read

$$T_{0,5} = T_{0,4} - \frac{1}{\eta_{\text{mech}}} \frac{C_p}{C_{pg}} \left(T_{0,3} - T_{0,2} \right).$$

From the definition of the isentropic efficiency for the turbine, it can be shown that

$$\frac{T_{0,5s}}{T_{0,4}} = 1 - \frac{1}{\eta_{\text{turbine}}}\left(1 - \frac{T_{0,5}}{T_{0,4}}\right).$$

Since the states 05 and 05s are on the same isentrope, it follows that

$$P_{0,5s} = P_{0,4}\left(\frac{T_{0,5s}}{T_{0,4}}\right)^{\frac{\gamma_g}{\gamma_g - 1}}.$$

If we substitute for T_{05s} from above, we get

$$P_{0,5} = P_{0,5s} = P_{0,4}\left[1 - \frac{1}{\eta_{\text{turbine}}}\left(1 - \frac{T_{0,5}}{T_{0,4}}\right)\right]^{\frac{\gamma_g}{\gamma_g - 1}}.$$

State 6

Since the LP turbine produces just enough work to run the fan, application of the energy equation 2.17 to the LP turbine and the fan gives

$$\dot{m}C_p\left(T_{0,2} - T_{0,1}\right) = \eta_{\text{mech}}(\dot{m}_H + \dot{m}_f)C_{pg}\left(T_{0,5} - T_{0,6}\right).$$

If we use the fact that $\dot{m}_f \ll \dot{m}_H$ and $\dot{m} = (B+1)\dot{m}_H$ then this equation can be simplified to read

$$T_{0,6} = T_{0,5} - \frac{1}{\eta_{\text{mech}}}(B+1)\frac{C_p}{C_{pg}}\left(T_{0,2} - T_{0,1}\right).$$

From the definition of the isentropic efficiency for the turbine, it can be shown that

$$\frac{T_{0,6s}}{T_{0,5}} = 1 - \frac{1}{\eta_{\text{turbine}}}\left(1 - \frac{T_{0,6}}{T_{0,5}}\right).$$

Since the states 06 and 06s are on the same isentrope, it follows that

$$P_{0,6s} = P_{0,5}\left(\frac{T_{0,6s}}{T_{0,5}}\right)^{\frac{\gamma_g}{\gamma_g - 1}}.$$

If we substitute for $T_{0,6s}$ from above, we get

$$P_{0,6} = P_{0,6s} = P_{0,5}\left[1 - \frac{1}{\eta_{\text{turbine}}}\left(1 - \frac{T_{0,6}}{T_{0,5}}\right)\right]^{\frac{\gamma_g}{\gamma_g - 1}}.$$

State (e, H)—Hot nozzle

The procedure for determining $P_{e,H}$, $T_{e,H}$ and $V_{e,H}$ is the same as that of the turbojet engine. Hence, only the final expressions for these quantities are given here.

As there is no heat or work interaction in the nozzle, $T_{0,e,H} = T_{0,6}$. The critical pressure ratio is given by (Eq. 7.17)

$$P_{e,H,\text{crit}} = P_{0,6}\left(1 - \frac{1}{\eta_{\text{nozzle}}}\frac{\gamma_g - 1}{\gamma_g + 1}\right)^{\frac{\gamma_g}{\gamma_g - 1}}$$

If the ambient pressure $P_a > P_{e,H,\text{crit}}$, then $P_{e,H} = P_a$. Otherwise, $P_{e,H} = P_{e,H,\text{crit}}$. Once $P_{e,H}$ is known, $T_{e,H}$ can be obtained from Eq. 7.18 as

$$T_{e,H} = T_{0,6}\left[1 - \eta_{\text{nozzle}}\left(1 - \left(\frac{P_{e,H}}{P_{0,6}}\right)^{\frac{\gamma_g - 1}{\gamma_g}}\right)\right].$$

The exit velocity can now be calculated in the same manner as before. Thus,

$$V_{e,H} = \sqrt{2C_{pg}(T_{0,e,H} - T_{e,H})}.$$

State (e, C)—Cold nozzle

Proceeding similarly, it is easy to see in this case that $T_{0,e,C} = T_{0,2}$ and the critical pressure ratio is given by

$$P_{e,C,\text{crit}} = P_{0,2}\left(1 - \frac{1}{\eta_{\text{nozzle}}}\frac{\gamma - 1}{\gamma + 1}\right)^{\frac{\gamma}{\gamma - 1}}.$$

If the ambient pressure $P_a > P_{e,C,\text{crit}}$, then $P_{e,C} = P_a$. Otherwise, $P_{e,C} = P_{e,C,\text{crit}}$. It follows from Eq. 7.18 that $T_{e,C}$ can be obtained as

$$T_{e,C} = T_{0,2}\left[1 - \eta_{\text{nozzle}}\left(1 - \left(\frac{P_{e,C}}{P_{0,2}}\right)^{\frac{\gamma - 1}{\gamma}}\right)\right].$$

The exit velocity can now be calculated in the same manner as before. Thus,

$$V_{e,C} = \sqrt{2C_p(T_{0,e,C} - T_{e,C})}.$$

The specific thrust can now be evaluated from Eq. 7.9.

7.8.2 TSFC

To evaluate thrust specific fuel consumption, we start by applying the energy equation 2.17 to the combustor,

$$\dot{m}_H C_p T_{0,3} = (\dot{m}_H + \dot{m}_f) C_{pg} T_{0,4} - \dot{m}_f H_{\text{fuel}}.$$

Upon rearranging this, we get

$$\dot{m}_f = \frac{\dot{m}_H \left(C_{pg} T_{0,4} - C_p T_{0,3} \right)}{H_{\text{fuel}} - C_{pg} T_{0,4}}.$$

This expression for the fuel mass flow rate is for the ideal case and so it should be written as

$$\dot{m}_{f,\text{ideal}} = \frac{\dot{m}_H \left(C_{pg} T_{0,4} - C_p T_{0,3} \right)}{H_{\text{fuel}} - C_{pg} T_{0,4}}.$$

Hence, from Eq. 7.12,

$$\dot{m}_{f,\text{actual}} = \frac{\dot{m}_{f,\text{ideal}}}{\eta_{\text{comb}}}.$$

Thrust specific fuel consumption can now be calculated from Eq. 7.4 using $\dot{m}_{f,\text{actual}}$.

Example 7.1 Consider the CF6-50C2 turbofan engine with the following specifications:

Bypass ratio	4.31
Fan pressure ratio	1.7
Overall pressure ratio	30.4
Air mass flow rate	670 kg/s
Turbine inlet temperature	1500 K

Assume

η_{inlet}	0.85
$\eta_{\text{fan}} = \eta_{\text{comp}}$	0.92
η_{turbine}	0.9
η_{nozzle}	0.97
η_{mech}	0.95 and
$\Delta P_{0,\text{comb}}$	5%.

For air, take $C_p = 1.005$ kJ/kg K and $\gamma = 1.4$. For the combustion gases, take $C_{pg} = 1.148$ kJ/kg K and $\gamma_g = 1.333$, $H_{\text{fuel}} = 45{,}000$ kJ/kg. Determine the thrust, TSFC and nozzle exit areas (cold and hot) for the following conditions:

1. Static ($V_a = 0$, $P_a = 100$ kPa and $T_a = 288$ K),
2. Take-off ($V_a = 60$ m/s, $P_a = 100$ kPa and $T_a = 288$ K) and
3. Cruise at 10 km ($M_a = 0.8$, $P_a = 26.5$ kPa and $T_a = 223$ K).

Solution:

(1) Static

Since $V_a = 0$, $P_{0,a} = P_a = 100$ kPa and $T_{0,a} = T_a = 288$ K.

Fan stream

At entry to fan, $T_{0,1} = T_{0,a} = 288$ K and

$$P_{0,1} = P_a\left[1 + \eta_{\text{inlet}}\left(\frac{T_{0,1}}{T_a} - 1\right)\right]^{\frac{\gamma}{\gamma-1}} = 100 \text{ kPa}.$$

At the fan exit, $P_{0,2} = \text{FPR } P_{0,1} = 170$ kPa and

$$T_{0,2} = T_{0,1}\left[1 + \frac{1}{\eta_{\text{fan}}}\left(\text{FPR}^{\frac{\gamma-1}{\gamma}} - 1\right)\right] = 339.2 \text{ K}.$$

For the fan nozzle,

$$P_{e,C,\text{crit}} = P_{0,2}\left(1 - \frac{1}{\eta_{\text{nozzle}}}\frac{\gamma-1}{\gamma+1}\right)^{\frac{\gamma}{\gamma-1}} = 87.88 \text{ kPa}.$$

Since $P_a > P_{e,C,\text{crit}}$, the fan nozzle is not choked. Hence

$$P_{e,C} = P_a = 100 \text{ kPa}$$

and

$$T_{e,C} = T_{0,2}\left[1 - \eta_{\text{nozzle}}\left(1 - \left(\frac{P_{e,C}}{P_{0,2}}\right)^{\frac{\gamma-1}{\gamma}}\right)\right] = 292.95 \text{ K}$$

and

$$V_{e,C} = \sqrt{2C_p(T_{0,e,C} - T_{e,C})} = 305.04 \text{ m/s}.$$

Therefore, fan thrust comes out to be

$$T_{\text{fan}} = \frac{B\dot{m}}{B+1}V_{e,C} = 166 \text{ kN}.$$

Core engine stream

At the compressor outlet

$$P_{0,3} = \text{PR } P_{0,1} = (\text{PR/FPR}) \; P_{0,2} = 3040 \, \text{kPa}$$

and

$$T_{0,3} = T_{0,2} \left[1 + \frac{1}{\eta_{\text{comp}}} \left(\left\{ \frac{\text{PR}}{\text{FPR}} \right\}^{\frac{\gamma-1}{\gamma}} - 1 \right) \right] = 811.05 \, \text{K}.$$

At the combustor exit

$$P_{0,4} = P_{0,3} \left(1 - \Delta P_{0,\text{comb}} \right) = 0.95 \; P_{0,3} = 2888 \, \text{kPa}$$

and $T_{04} = 1500$ K. At the HP turbine exit

$$T_{0,5} = T_{0,4} - \frac{1}{\eta_{\text{mech}}} \frac{C_p}{C_{pg}} \left(T_{0,3} - T_{0,2} \right) = 1065.23 \, \text{K}$$

and

$$P_{0,5} = P_{0,4} \left[1 - \frac{1}{\eta_{\text{turbine}}} \left(1 - \frac{T_{0,5}}{T_{0,4}} \right) \right]^{\frac{\gamma_g}{\gamma_g - 1}} = 609.36 \, \text{kPa}.$$

At the LP turbine exit

$$T_{0,6} = T_{0,5} - \frac{1}{\eta_{\text{mech}}} (B + 1) \frac{C_p}{C_{pg}} \left(T_{0,2} - T_{0,1} \right) = 814.5 \, \text{K}$$

and

$$P_{0,6} = P_{0,5} \left[1 - \frac{1}{\eta_{\text{turbine}}} \left(1 - \frac{T_{0,6}}{T_{0,5}} \right) \right]^{\frac{\gamma_g}{\gamma_g - 1}} = 181.02 \, \text{kPa}.$$

At the hot nozzle exit,

$$P_{e,H,\text{crit}} = P_{0,6} \left(1 - \frac{1}{\eta_{\text{nozzle}}} \frac{\gamma_g - 1}{\gamma_g + 1} \right)^{\frac{\gamma_g}{\gamma_g - 1}} = 95.72 \, \text{kPa}.$$

Since $P_a > P_{e,H,\text{crit}}$, the hot nozzle is not choked. Hence

$$P_{e,H} = P_a = 100 \, \text{kPa}$$

and

$$T_{e,H} = T_{0,6} \left[1 - \eta_{\text{nozzle}} \left(1 - \left(\frac{P_{e,H}}{P_{0,6}} \right)^{\frac{\gamma_g - 1}{\gamma_g}} \right) \right] = 705.6 \, \text{K}.$$

The exit velocity can now be calculated in the same manner as before. Thus,

$$V_{e,H} = \sqrt{2C_{pg}(T_{0,e,H} - T_{e,H})} = 500\,\text{m/s}$$

Core engine thrust can now be evaluated.

$$T_{core} = \frac{\dot{m}}{B+1}V_{e,H} = 63\,\text{kN}.$$

Therefore the total thrust

$$T = 166\,\text{kN} + 63\,\text{kN} = 229\,\text{kN}.$$

This compares well with the quoted value of 233.6 kN (see Table B.3 in *Elements of Gas Turbine Propulsion* by Mattingly). It is important to note that the fan generates nearly 73% of the overall thrust as this is a high bypass ratio turbofan engine.
   The fuel flow rate can be evaluated from

$$\dot{m}_f = \frac{\dot{m}_H\,(C_{pg}T_{0,4} - C_p T_{0,3})}{H_{\text{fuel}} - C_{pg}T_{0,4}} = 2.64\,\text{kg/s}.$$

Therefore

$$TSFC = \frac{\dot{m}_f}{T} = 0.0414\,\text{kg/hN}.$$

This compares reasonably well with the quoted value of 0.03781 kg/h N.

(2) Take-off

Although $V_a = 60$ m/s, $V_a^2/2C_p \approx 1.8$ K. So, we can assume P_{0a} and T_{0a} to be the same as in the previous case without any significant loss of accuracy. Therefore, the fan thrust comes out to be

$$T_{fan} = \frac{B\dot{m}}{B+1}(V_{e,C} - V_a) = 133.35\,\text{kN}.$$

Similarly, the core engine thrust can be calculated from

$$T_{core} = \frac{\dot{m}}{B+1}(V_{e,H} - V_a) = 55.44\,\text{kN}.$$

Therefore the total thrust

$$T = 133.35\,\text{kN} + 55.44\,\text{kN} = 188.79\,\text{kN}.$$

The 17% reduction in total thrust due to the intake momentum drag should be noted.
   Fuel flow rate remains the same at 2.64 kg/s. Therefore

$$TSFC = \frac{\dot{m}_f}{T} = 0.05\,\text{kg/hN}.$$

(3) Cruise

From the given cruise condition, we can get

$$P_{0,a} = P_a\left(1 + \frac{\gamma - 1}{2}M_a^2\right)^{\frac{\gamma}{\gamma-1}} = 40.4\,\text{kPa}$$

and

$$T_{0,a} = T_a\left(1 + \frac{\gamma - 1}{2}M_a^2\right) = 251.5\,\text{K}$$

and

$$V_a = M_a\sqrt{\gamma R T_a} = M_a\sqrt{(\gamma - 1)C_p T_a} = 239.5\,\text{m/s.}$$

Fan stream

At entry to fan, $T_{0,1} = T_{0,a} = 251.5$ K and

$$P_{0,1} = P_a\left[1 + \eta_{inlet}\left(\frac{T_{01}}{T_a} - 1\right)\right]^{\frac{\gamma}{\gamma-1}} = 38.04\,\text{kPa.}$$

At the fan exit, $P_{0,2} = \text{FPR}\,P_{0,1} = 64.67$ kPa and

$$T_{0,2} = T_{0,1}\left[1 + \frac{1}{\eta_{fan}}\left(\text{FPR}^{\frac{\gamma-1}{\gamma}} - 1\right)\right] = 296.30\,\text{K.}$$

For the fan nozzle,

$$P_{e,C,\text{crit}} = P_{0,2}\left(1 - \frac{1}{\eta_{nozzle}}\frac{\gamma - 1}{\gamma + 1}\right)^{\frac{\gamma}{\gamma-1}} = 33.43\,\text{kPa.}$$

Since $P_a < P_{e,C,\text{crit}}$, the fan nozzle is choked. Hence

$$P_{e,C} = 33.43\,\text{kPa}$$

and

$$T_{e,C} = T_{0,2}\left[1 - \eta_{nozzle}\left(1 - \left(\frac{P_{e,C}}{P_{0,2}}\right)^{\frac{\gamma-1}{\gamma}}\right)\right] = \frac{2}{\gamma + 1}T_{02} = 246.92\,\text{K}$$

and

$$V_{e,C} = \sqrt{2C_p(T_{0,e,C} - T_{e,C})} = 315.016\,\text{m/s.}$$

Therefore, fan thrust comes out to be

$$T_{\text{fan}} = \frac{B\dot{m}}{B+1}\left(V_{e,C} - V_a + \frac{RT_{e,C}}{P_{e,C}V_{e,C}}(P_{e,C} - P_a)\right)$$

$$T_{\text{fan}} = 41.075\,\text{kN} + 25.365\,\text{kN} = 66.44\,\text{kN}.$$

Core engine stream

At the compressor outlet

$$P_{0,3} = \text{PR}\ P_{0,1} = (\text{PR/FPR})\ P_{0,2} = 1156.4\,\text{kPa}$$

and

$$T_{0,3} = T_{0,2}\left[1 + \frac{1}{\eta_{\text{comp}}}\left(\left\{\frac{\text{PR}}{\text{FPR}}\right\}^{\frac{\gamma-1}{\gamma}} - 1\right)\right] = 708.4\,\text{K}.$$

At the combustor exit $T_{0,4} = 1500$ K and

$$P_{0,4} = P_{0,3}\left(1 - \Delta P_{0,\text{comb}}\right) = 0.95\ P_{0,3} = 1098.6\,\text{kPa}.$$

At the HP turbine exit

$$T_{0,5} = T_{0,4} - \frac{1}{\eta_{\text{mech}}}\frac{C_p}{C_{pg}}\left(T_{0,3} - T_{0,2}\right) = 1120.26\,\text{K}$$

and

$$P_{0,5} = P_{0,4}\left[1 - \frac{1}{\eta_{\text{turbine}}}\left(1 - \frac{T_{0,5}}{T_{0,4}}\right)\right]^{\frac{\gamma_g}{\gamma_g - 1}} = 292.83\,\text{kPa}.$$

At the LP turbine exit

$$T_{0,6} = T_{0,5} - \frac{1}{\eta_{\text{mech}}}(B + 1)\frac{C_p}{C_{pg}}\left(T_{0,2} - T_{0,1}\right) = 901.24\,\text{K}$$

and

$$P_{0,6} = P_{0,5}\left[1 - \frac{1}{\eta_{\text{turbine}}}\left(1 - \frac{T_{0,6}}{T_{0,5}}\right)\right]^{\frac{\gamma_g}{\gamma_g - 1}} = 109.86\,\text{kPa}.$$

At the hot nozzle exit,

$$P_{e,H,\text{crit}} = P_{0,6}\left(1 - \frac{1}{\eta_{\text{nozzle}}}\frac{\gamma_g - 1}{\gamma_g + 1}\right)^{\frac{\gamma_g}{\gamma_g - 1}} = 64.04\,\text{kPa}.$$

Since $P_a < P_{e,H,\text{crit}}$, the hot nozzle is choked. Hence

$$P_{e,H} = P_{e,H,\text{crit}} = 64.04 \text{ kPa}$$

and

$$T_{e,H} = T_{0,6} \left[1 - \eta_{\text{nozzle}} \left(1 - \left(\frac{P_{e,H}}{P_{0,6}} \right)^{\frac{\gamma_g - 1}{\gamma_g}} \right) \right] = \frac{2}{\gamma_g + 1} T_{0,6} = 772.61 \text{ K}.$$

The exit velocity can now be calculated in the same manner as before. Thus,

$$V_{e,H} = \sqrt{2C_{pg}(T_{0,e,H} - T_{e,H})} = 543.47 \text{ m/s}.$$

Core engine thrust can now be evaluated.

$$T_{\text{core}} = \frac{\dot{m}}{B+1} \left(V_{e,H} - V_a + \frac{R_g T_{e,H}}{P_{e,H} V_{e,H}} (P_{e,H} - P_a) \right)$$

$$T_{\text{core}} = 38.35 \text{ kN} + 28 \text{ kN} = 66.35 \text{ kN}.$$

Therefore, the total thrust

$$T = 66.44 \text{ kN} + 66.35 \text{ kN} = 132.79 \text{ kN}.$$

Note that there is an overall reduction in the thrust at the higher altitude. However, this is not due to a decrease in mass flow rate, since it has been kept the same. The decrease in the fan thrust is entirely due to the decrease in the density of the incoming air. For the same reason, the work required by the fan as well as the compressor are lower at the higher altitude. Consequently, the energy extracted by the turbines is less, resulting in a higher core engine thrust.

The fuel flow rate can be evaluated from

$$\dot{m}_f = \frac{\dot{m}_H \left(C_{pg} T_{0,4} - C_p T_{0,3} \right)}{H_{\text{fuel}} - C_{pg} T_{0,4}} = 2.94 \text{ kg/s}$$

$$\text{TSFC} = \frac{\dot{m}_f}{T} = 0.08 \text{ kg/hN}.$$

These values of cruise thrust and TSFC compare very poorly with the quoted values of 51.41 kN and 0.0642 kg/h N, respectively (see Table B.3 in *Elements of Gas Turbine Propulsion* by Mattingly). The reason for this discrepancy is examined next.

For the sea level static condition, the mass flow rate through the cold nozzle is equal to

$$\dot{m}_C = \frac{B}{B+1} \dot{m} = 543.823 \text{ kg/s}.$$

Fig. 7.12 Core engine (hot) nozzle and fan (cold) nozzle in a turbofan engine. *Courtesy* Rolls-Royce plc

IBERIA

Since $\dot{m}_C = \rho_{e,C} V_{e,C} A_{e,C} = P_{e,C}/(RT_{e,C}) V_{e,C} A_{e,C}$, using the values calculated above, we can get the cold nozzle area as $1.5\,\mathrm{m}^2$. The mass flow rate through the hot nozzle is equal to

$$\dot{m}_H = \frac{1}{B+1}\dot{m} = 126.177\,\mathrm{kg/s}.$$

Since $\dot{m}_H = P_{e,H}/(R_g T_{e,H}) V_{e,H} A_{e,H}$, the hot nozzle exit area can be evaluated as $0.511\,\mathrm{m}^2$. Since the predicted values of thrust and TSFC for the sea level static condition agree well with the quoted values, we can safely assume that the calculated nozzle areas are correct.

Proceeding similarly for the cruise condition, the required cold and hot nozzle exit areas for the given mass flow rate come out to be $3.661\,\mathrm{m}^2$ and $0.8855\,\mathrm{m}^2$, respectively. These are different from the corresponding values for the static condition. Since the nozzle areas are fixed and usually cannot be changed in commercial aircraft engines (Fig. 7.12), it is clear that this is the reason for the discrepancy in the calculated thrust and TSFC values. This calculation serves as a good illustrative example for the thrust and TSFC calculation procedure and sheds some light on the interplay between the different terms in the thrust equation under different operating conditions. It also illustrates the need to change operating conditions with altitude. In an actual engine, a new operating condition that generates the required thrust at a given altitude has to be determined. The fan pressure ratio, overall pressure ratio and fuel mass flow rate have to be adjusted until a new stable operating condition is achieved.[11] This is demonstrated next for the problem under consideration. For the sake of simplicity, the mass flow rate of air and fuel alone are adjusted, while the

---

[11] Although the bypass ratio depends on the dimensions of the fan and the core engine for a given altitude and flight speed, once the cold and hot nozzles choke, the mass flow rate that can be swallowed by the fan and the core engine, and hence the bypass ratio, are determined by the nozzles.

other parameters are kept constant at their design operating values (sea level static condition).

Since the calculations require the cold nozzle area to be higher by a factor of $3.661\,\text{m}^2/1.5\,\text{m}^2 = 2.441$, if the cruise air mass flow rate were reduced by the same factor, then the cold nozzle area can remain the same. Hence, the cruise air mass flow rate has to be changed from 670 to $670/2.441 = 274.48$ kg/s. The corresponding mass flow rate through the hot nozzle is $274.48/(4.31 + 1) = 51.69$ kg/s, which implies that the hot nozzle area required now is $0.363\,\text{m}^2$. This is less than the desired value of $0.511\,\text{m}^2$. It is not possible to use the same strategy as before, i.e., adjusting the mass flow rate through the hot nozzle, as this would change the mass flow rate through the cold nozzle also (with bypass ratio remaining the same). Since the hot nozzle is always choked at such altitudes, the mass flow rate $\dot{m}_H \propto A_{e,H} P_{0,6}/\sqrt{T_{0,6}}$. It can be seen from this expression that in order to pass a given mass flow rate through a hot nozzle with a fixed exit area, $P_{0,6}$ and $T_{0,6}$, either individually or together, have to be adjusted. For the problem under consideration, the matching of the mass flow rate to the nozzle area can be accomplished by decreasing the turbine inlet temperature or, equivalently, the fuel mass flow rate. It is important to note that altering the turbine inlet temperature does not affect the fan stream at all. If the turbine inlet temperature were reduced to 1313.8 K (this has to be determined iteratively), the required hot nozzle exit area comes out to be $0.510864\,\text{m}^2$, which is close enough to the desired value. This requires a reduction in fuel mass flow rate from 2.94 to 0.945 kg/s.

Once the mass flow rate and the turbine inlet temperature are determined, the calculation procedure is the same as before. The fan thrust is now equal to 27.2 kN (16.8 kN + 10.4 kN), and the core engine thrust is equal to 17.8 kN (12.6 kN + 5.2 kN). The total thrust is 45 kN and the TSFC is 0.0756 kg/h N. These values compare quite well with the values quoted for this engine. Of course, it must be borne in mind that we have assumed the values for the parameters as well as the component efficiencies to be the same at the off-design operating condition as well. In reality, this is not the case.

The procedure described above for calculating the thrust matches the mass flow rate as well as the work for each compressor/turbine pair but **not the RPM**. In this sense, the calculation procedure may be considered to be only thermodynamic in nature. Matching of the RPM, in addition, will make the procedure aerodynamic as well, since this requires the compressor and turbine maps—operating characteristics depicting variation of the mass flow rate through the component and the efficiency with pressure ratio and RPM. Hence, the determination of a stable off-design operating point is more involved than the simple procedure described above. The aero-thermodynamic component matching procedure will yield operating points, which meet the requirements of all the components and generate the required thrust. The interested reader is referred to the book by Saravanamuttoo, Rogers and Cohen for detailed off-design component matching procedure for single spool and two-spool engines.

Exercises

For all the problems, relevant ambient conditions, property data and component efficiencies may be taken from Worked Example 7.1. Wherever possible, engine specifications have been taken from Appendix B in *Elements of Gas Turbine Propulsion* by Mattingly. Readers are encouraged to compare the answers given here with the corresponding values given there.

7.1 The J-75-P-17 military turbojet engine has the following specifications:

Overall pressure ratio	12
Air mass flow rate	114 kg/s
Turbine inlet temperature	1150 K

Determine the thrust and the TSFC under static, sea level conditions.
[76.2 kN, 0.08712 kg/h N]

7.2 The J-57-P-23 military turbojet engine with a convergent nozzle has the following specifications:

Overall pressure ratio	11.5
Air mass flow rate	75 kg/s
Turbine inlet temperature	1150 K
Afterburner exit temperature	1666 K
Stagnation pressure loss in afterburner	10%
(during operation)	

Determine the thrust and TSFC under static, sea level conditions without and with afterburner operation. Also, determine the the nozzle exit area for both cases.
[50.25 kN, 0.08784 kg/h N, 0.194 m^2, 67 kN, 0.16416 kg/h N, 0.3 m^2]

7.3 The Pratt and Whitney JT9D-59A commercial turbofan engine has the following specifications:

Bypass ratio	4.9
Fan pressure ratio	1.54
Overall pressure ratio	24.5
Air mass flow rate	743.3 kg/s
Air-fuel ratio	55.3:1
(based on core engine air mass flow rate)	

Calculate the thrust and TSFC under sea level static conditions.

[215.2 kN, 0.03816 kg/h N]

7.4 The Rolls-Royce RB211-882 commercial turbofan engine has the following specifications:

Bypass ratio	6
Fan pressure ratio (static)	1.5
Fan pressure ratio (cruise)	1.55
Overall pressure ratio	39

Calculate the thrust, TSFC and nozzle exit areas for the following conditions:

1. Static (Turbine inlet temperature = 1550 K, Air mass flow rate = 1200 kg/s) and
2. Cruise at 10 km (M_a = 0.83, Turbine inlet temperature = 1434 K, Air mass flow rate = 535 kg/s).

[356.9 kN, 0.036 kg/h N, 3.266 m², 0.681 m², 74.28 kN, 0.07452 kg/h N, 3.266 m², 0.681 m²]

7.5 The General Electric GE90-B4 commercial turbofan engine has the following specifications:

Bypass ratio	8.4
Fan pressure ratio (static)	1.58
Fan pressure ratio (cruise)	1.65
Overall pressure ratio	39.3

Calculate the thrust, TSFC and nozzle exit areas for the following conditions:

1. Static (Turbine inlet temperature = 1650 K, Air mass flow rate = 1340 kg/s) and
2. Cruise at 10 km (M_a = 0.8, Turbine inlet temperature = 1448 K, Air mass flow rate = 578.5 kg/s).

[373.3 kN, 0.0324 kg/h N, 3.57 m², 1.19 m², 65.4 kN, 0.06984 kg/h N, 3.57 m², 1.19 m²]

7.6 The CFM56-5C commercial turbofan engine has the following specifications:

Bypass ratio	6.6
Fan pressure ratio	1.5
Overall pressure ratio	31.5

Calculate

1. the static thrust and TSFC under sea level conditions (Turbine inlet temperature = 1550 K, Air mass flow rate = 466 kg/s) and
2. the air mass flow rate, turbine inlet temperature, thrust and TSFC for cruise at $M_a = 0.8$ and 10 km altitude.

[136.6 kN, 0.035784 kg/h N, 197.7 kg/s, 1372 K, 25.06 kN, 0.0738 kg/h N]

Chapter 8
Ramjet and Scramjet Engine

The turbojet and turbofan engines discussed in detail in the previous chapters are designed for high subsonic cruise speeds. The addition of an afterburner allows the turbojet engine to operate at transonic speeds for short periods of time. As shown in Fig. 1.1, for sustained supersonic flight speeds, ramjet or scramjet engines are necessary. These are discussed in this chapter. Keeping the intended audience in mind, only the basic concepts and ideas are presented here. The interested reader should consult the propulsion books given in the list of suggested reading for further details.

8.1 Ramjet Engine

The ramjet engine consists of an inlet, combustor and a nozzle. A schematic diagram of a ramjet is shown in Fig. 2.5. Air enters the inlet where it is compressed by a series of oblique shocks usually terminating in a normal shock and then enters the combustor where it is mixed with the fuel and burned. The hot gases are then expelled through the nozzle thereby developing thrust. As the compression of the air is achieved through deceleration in the inlet, the cruise speed needs to be high for this design to be viable. Ramjets typically operate in the 2–4 cruise Mach number range. However, the flow is subsonic at entry to the combustor.

In comparison with the turbojet and turbofan engines discussed in the previous chapters, the following points about the ramjet engine should be noted.

- Since the compression of the air is achieved entirely in the intake, the compressor and hence the turbine are not required. Thus, the ramjet engine does not have any moving components. In view of the issues discussed in Chaps. 5 and 6 in connection with the design of fan, compressor and turbine blades, the absence of any moving parts is definitely highly desirable. However, the absence of a compressor also means that the intake can operate effectively only after supersonic flight speeds are attained.

- The combustor is very simple. In fact, the combustor design used in the ramjet engine is the same as the one used in afterburners (Fig. 5.11). Since the turbine and associated complications such as the cooling of the highly stressed blades are absent, the peak temperature in ramjet engines is usually much higher than that seen in turbojet and turbofan engines.
- The fixed throat nozzle is also very simple in design and lacks any additional mechanical complexities such as thrust reversing mechanism. It is convergent or convergent divergent in design depending upon the stagnation pressure at entry.
- In contrast to the last two points, the inlet, which is usually very simple in design in the conventional engines, is quite complicated in the case of a ramjet engine.

In view of the points mentioned above, it is clear that the intake design deserves the most attention, and we turn to this next.

8.2   Supersonic Intakes

The main function of the intake[1] is to achieve compression of the captured free stream air through deceleration. In addition, it should

- capture the required mass flow rate of air
- achieve the compression with minimal loss of stagnation pressure
- deliver the compressed air to the combustor with minimal flow distortion
- operate in a stable manner.

These issues are discussed later in this section. First, we take a close look at the manner in which the deceleration (and hence compression) of the air is brought about. Deceleration of the supersonic freestream can be achieved either using an appropriately shaped passage (a diffuser) or through a combination of oblique and or normal shocks. Intakes which employ the former strategy are called internal compression intakes and those that utilize the latter strategy are called external compression intakes. Intakes that are designed to use both these strategies are called, appropriately enough, mixed compression intakes.

---

[1] In the literature, the terms intake and inlet are used interchangeably. The former name is British in origin, while the latter is American. The interested reader should read the preface of the book *Intake Aerodynamics* by Seddon and Goldsmith, for a rather nice exposition of this point.

Fig. 8.1 Illustration of a
normal shock diffuser.
Capture streamtube is
indicated by dashed lines

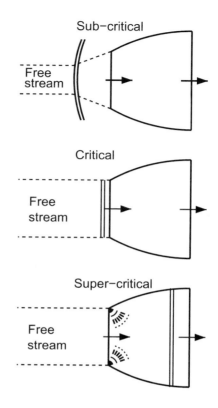

8.2.1 Internal Compression Intakes

8.2.1.1 Normal Shock Diffuser

The normal shock diffuser is the simplest of all the internal compression intake
designs. The diffuser is a diverging passage and hence is intended to diffuse a subsonic flow. Consequently, the supersonic freestream is decelerated to subsonic Mach
number across a normal shock wave before entry into the diffuser (Fig. 8.1). At any
freestream Mach number, depending upon the mass flow rate requirements of the
engine (or the back pressure), the diffuser operates in any one of three modes, namely,
critical, sub-critical and super-critical.

When the mass flow rate required by the engine just equals the maximum mass
flow rate that the intake can capture, then the intake operates in the critical mode.
The normal shock in this case, stands just at the entry to the diffuser and the intake
capture area, is the highest possible.

In the sub-critical mode of operation, the mass flow rate required by the engine
is less than the critical value, or, equivalently, the back pressure is higher than the
critical value. Some of the air that would have otherwise entered the intake has to

be spilled (or diverted) now. This is accomplished with the shock standing some distance away in front of the intake, instead of at the entry to the diffuser. In addition, this shock is not a normal shock as in the case of the critical mode, but is a bow shock instead.[2] Hence, the stagnation pressure recovery in the sub-critical mode will be somewhat better than that of the critical mode. As Fig. 8.1 shows, the capture area is less now, consistent with the reduced mass flow rate requirement of the engine. The diffusion of the captured air after passage through the shock and before entering the diffuser can be seen from the diverging capture streamtube. Consequently, the static pressure at the diffuser exit will be high enough to match the increased back pressure.

In the super-critical mode of operation, the mass flow rate required by the engine is more than the critical value, or, equivalently, the back pressure is less than the critical value. Since the capture mass flow rate is already at the maximum possible value in the critical mode itself, the intake cannot meet this increased requirement. However, the net effect of the reduced back pressure is to suck the normal shock into the diffuser. The supersonic freestream flow is *accelerated* in the diverging passage, as seen from the expansion fans shown in the diffuser for super-critical mode of operation in Fig. 8.1. Consequently, the normal shock occurs at a higher Mach number than before with an attendant increased loss of stagnation pressure. The normal shock locates itself to match the back pressure.

The stagnation pressure loss across the normal shock (in any mode of operation) is prohibitively high (30% or more) beyond a freestream Mach number of 2. In practice, the pressure rise across the normal shock even at a Mach number less than 2 will be high enough to trigger the separation of the boundary layer on the walls of the diffuser. Thus, the normal shock diffuser cannot be used for freesream Mach numbers greater than 1.5 or so.

8.2.1.2   Converging–Diverging Diffuser

The main disadvantage of the normal shock diffuser is that even at design (or critical) operating condition at any Mach number, the loss of stagnation pressure is quite high. This can be addressed by designing the passage for shock free operation at a design Mach number M_d. The converging–diverging diffuser shown in Fig. 8.2 is an example of such an intake. Here, the intake passage cross-section is determined from area Mach number relationship for an isentropic flow. At the design operating condition, a weak normal shock is located just downstream of the throat and the throat Mach number is just above 1 (Fig. 8.2). The flow is supersonic in the converging passage and subsonic in the diverging passage. The capture area is equal to the area of the intake at the inlet. Although conceptually very simple (since it is just a CD nozzle in reverse), this type of fixed throat internal compression intake is plagued by starting

---

[2] Across a normal shock, the flow turning angle is zero, whereas it is non-zero across a bow shock. This allows the excess air to be deflected and spilled around the intake. See Chap. 7 in Fundamentals of Gas Dynamics by V. Babu for a discussion of the bow shock structure.

Fig. 8.2 Illustration of a converging–diverging diffuser. Capture streamtube is indicated by dashed lines

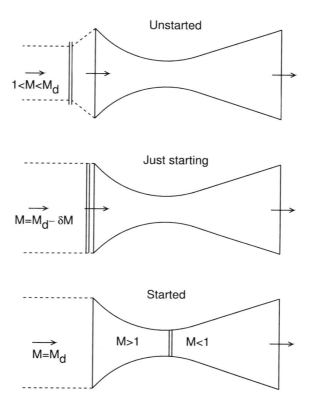

problem and stability problem. These are briefly outlined next, and the reader is referred to *Mechanics and Thermodynamics of Propulsion* by Hill and Peterson for a detailed discussion of the sequence of events preceding the successful starting of such an intake.

During start-up, when the freestream Mach number is supersonic but less than the design Mach number, there will be a normal shock standing in front of the intake some distance away (Fig. 8.2). The stand-off distance depends on the actual freestream Mach number. The flow downstream of the normal shock and in the entire diffuser is subsonic (except at the throat where the Mach number can be sonic). The capture area is also less than the inlet area. The intake in this case is said to be *unstarted*. As the freestream Mach number increases, the normal shock in front of the intake moves closer. When the freestream Mach number equals the design Mach number, the normal shock stands just at the entrance to the intake (Fig. 8.2). At this point, the intake is on the verge of starting. If the freestream Mach number is increased slightly, then the normal shock is swallowed and moves into the converging part of the diffuser, which is an unstable location. The freestream Mach number can now be decreased to the design value and the normal shock will then position itself just downstream of the throat, which is where it is most stable. In this mode, wherein the entire flow is shock free, the intake is said to be *started*. Any slight disturbance of the freestream conditions causes the swallowed shock to be disgorged and the entire start-up process including the over-speeding of the intake has to be repeated, in order to get the intake started again.[3]

The start-up difficulty of internal compression intakes in general can be traced to the mismatch between the mass flow that the intake is designed to pass and the mass flow that tries to enter the intake (which is usually more than the design value). Left to itself, the intake adjusts by causing a shock to stand in front and spill some of the flow. The mismatch can be handled more gracefully by either having an adjustable throat or bleed ports on the walls of the diffuser. In the first design, the throat area is adjusted to accept the mass flow rate that enters the intake. In the second design, the excess mass flow rate is spilled in a controlled manner through the bleed ports. In both cases, the intake is shock free and stable, albeit being mechanically more complex.

8.2.2 External Compression Intakes

External compression intakes achieve deceleration and compression of the freestream through a sequence of oblique shocks generated from external ramps, usually terminating in a normal shock. The air is then delivered to the combustor through an internal passage formed by a cowl and the ramp structure as shown in Fig. 8.3. Under design operating condition, the ramps can be designed to have either an unfocussed or a focussed system of shocks with the normal shock standing at the cowl passage entrance. Both these are illustrated in Fig. 8.3. As shown in this figure, for a given axial position of the ramp structure, the capture area is smaller when the shocks are unfocussed. The capture area can be increased to the same value as the design in Fig. 8.3b, by moving the ramp structure forward (the readers should demonstrate this on their own). However, since the ramp structure is now longer, there will be an increase in friction drag arising from the boundary layer on the ramp surface. Moreover, it can be seen from Fig. 8.3a that, when the shocks are not focussed on the cowl lip but extend beyond, the resulting pressure on the external surface of the cowl is higher. Consequently, the pressure drag is also higher (this is discussed in more detail in the next subsection). For these reasons, external compression intakes are designed to operate as shown in Fig. 8.3b. The oblique shocks from the ramps impinge on the cowl lip. The lip interior surface is designed so that there are no reflected shocks and the shock train is terminated by a normal shock at the cowl lip. The external compression intake can also operate in critical, sub-critical or super-critical mode. These modes of operation and the total pressure recovery and capture mass flow rate associated with them are discussed next.

---

[3] It should be noted that, unlike the normal shock diffuser, the converging–diverging diffuser, does not have critical, sub-critical or super-critical modes of operation. It operates in one of two modes—started or unstarted.

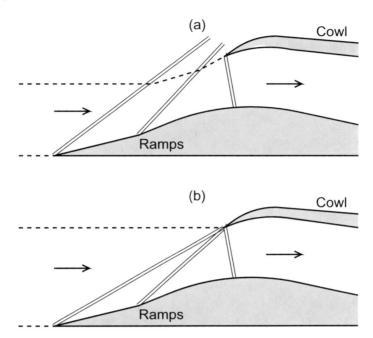

Fig. 8.3 Illustration of external compression intakes. **a** Shocks unfocussed. **b** Shocks focussed on the cowl lip. Capture streamtube is indicated by dashed lines. Adapted from Intake Aerodynamics by Seddon and Goldsmith

8.2.3 Intake Characteristic Curve

Figure 8.4 illustrates the three modes of operation of the external compression intake. Not surprisingly, the observations made earlier for the normal shock internal compression intake operating in the critical, sub-critical and super-critical modes are applicable here as well.

When the mass flow rate required by the engine just equals the maximum mass flow rate that the intake can capture, then the intake operates in the critical mode, which is usually the design condition. The oblique shocks from the external ramps converge on the cowl lip and the terminal normal shock stands just at the entry to the cowl passage and the intake capture area is the highest possible.

In the sub-critical mode of operation, the mass flow rate required by the engine is less than the critical value, or, equivalently, the back pressure is higher than the critical value. Such a condition downstream of the cowl passage forces the terminal normal shock to move out and stand some distance away in front of the intake, instead of at the cowl lip, as shown in Fig. 8.4. The oblique shocks from the ramps are no longer focussed on the cowl lip but move outward and away from the cowl. This causes flow deflection away from and around the cowl and consequently the capture area decreases, thereby matching the reduced mass flow rate requirement of the engine.

Fig. 8.4 Illustration of different modes of operation of an external compression intake

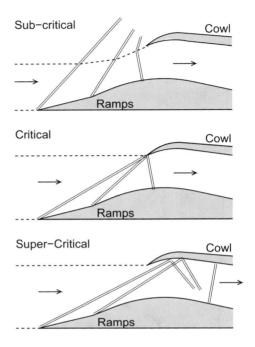

In addition, the terminal shock is not entirely a normal shock as in the case of the critical mode, but is a bow shock instead. Hence, the stagnation pressure recovery in the sub-critical mode will be slightly better than that of the critical mode. Once again, the diffusion of the captured air after passage through the oblique shocks and before entering the cowl passage can be seen from the diverging capture streamtube. Consequently, the static pressure at the end of the cowl passage will be high enough to match the increased back pressure.

In the super-critical mode of operation, the mass flow rate required by the engine is more than the critical value, or, equivalently, the back pressure is less than the critical value. Since the freestream is supersonic, the capture area cannot be increased beyond its value in the critical mode and so the capture mass flow rate cannot be increased beyond the critical mass flow rate. However, the reduced back pressure causes the normal shock to move inside the cowl passage. The oblique shocks from the ramps are unfocussed and impinge on the cowl inside surface and are subsequently reflected. Consequently, the stagnation pressure loss is higher in the super-critical mode. The normal shock locates itself to match the back pressure.

A major drawback of the sub-critical mode of operation is the so called *wave drag* or *shock drag*. It can be seen from Figs. 8.1 and 8.4 that the pressure on the exterior surface of the cowl (or diffuser) is higher in the sub-critical mode when compared to the critical mode, owing to the passage of the spilled flow. Consequently, the pressure drag is also higher. This drag force, which arises entirely from the spillage of the flow around the intake is called *wave drag* or *shock drag*. At Mach numbers beyond 2, the

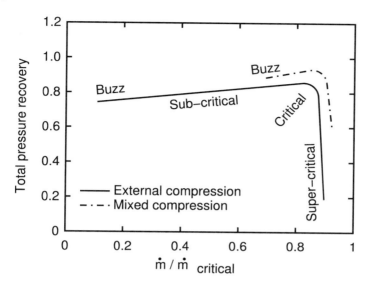

Fig. 8.5 Illustration of the characteristic curve for external and mixed compression intakes. Adapted from Elements of Gas Turbine Propulsion by Mattingly

shock drag can be quite high during sub-critical mode of operation and hence this mode should be avoided.

Trends in capture mass flow rate and stagnation pressure recovery discussed above are illustrated in Fig. 8.5. The resulting curve is called the intake characteristic curve. The steep loss in the stagnation pressure recovery in super-critical mode is evident from this figure. Although this figure shows that there is no significant deterioration in the stagnation pressure recovery during sub-critical mode of operation, the increased external drag force due to wave drag cannot be discerned from this figure.

As mentioned at the beginning of this section, in addition to stagnation pressure recovery and mass flow rate, flow distortion is an important performance metric of an intake. Flow distortion at an axial station is defined as

$$\text{Flow distortion} = \frac{P_{0,max} - P_{0,min}}{P_{0,mean}} \qquad (8.1)$$

where $P_{0,max}$, $P_{0,min}$ and $P_{0,mean}$ are, respectively, the maximum, minimum and mean value of the stagnation pressure at the given axial station. This definition for flow distortion is thus a measure of the spatial non-uniformity of the flow. Ideally, it should be zero, but in reality, non-uniformity is introduced primarily by shocks which tend to compress the flow primarily in one direction as well as by boundary layers.

8.2.4 Mixed Compression Intake

Mixed compression intake design addresses disadvantages in internal and external compression intakes, namely, loss of stagnation pressure due to boundary layers in the former and high external drag in the latter. The idea here is design the cowl passage in an external compression intake as an internal compression diffuser and compress the freestream air partly through oblique shocks and partly through diffusion. This reduces the compression to be achieved in each, thereby mitigating the disadvantages mentioned above.

Figure 8.6 illustrates two mixed compression intake designs. In both designs, four shocks are utilized. In the first design, two shocks are generated from the external ramps and the remaining two are inside the cowl passage, and is aptly called 50/50 design. In the second design, which is called the 25/75 design, one shock is an external shock with the remaining three being internal shocks. In both designs, the terminal shock is positioned just downstream of the throat (similar to that of a converging–diverging diffuser). The contrast in the cowl interior surface between the mixed (Fig. 8.6) and external compression intake (Fig. 8.3) is noteworthy. In the mixed compression intake, the cowl interior surface is designed to either reflect incident shocks (50/50 design) or generate shocks from compression corners (25/75 design).

The operating curve for an external compression intake is shown in Fig. 8.5. Although mixed compression intakes perform better than external compression intakes, their range of stable operation is narrower. If the operating conditions, namely, flight Mach number, mass flow rate and/or back pressure conditions can be maintained within this narrow range, then mixed compression intakes are more desirable.

Fig. 8.6 Illustration of mixed compression intakes. Capture streamtube is indicated by dashed lines. Adapted from Intake Aerodynamics by Seddon and Goldsmith

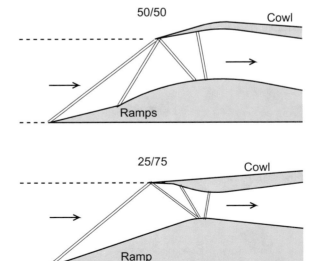

8.2.5 *Stability of Intake Operation*

In our discussion so far, we have ignored the effect of viscosity, particularly as it pertains to the presence of boundary layers on the solid surfaces. The flow fields that we have described have all been obtained based on gas dynamic considerations alone. In other words, boundary layer thickness has been assumed to be zero. In reality, boundary layers are present on the solid surfaces and they do exert considerable influence on the overall flow field. Two important aspects in this connection are

- Shock—boundary layer interaction
  Whenever a shock impinges on a solid surface, the pressure rise across the shock causes the boundary layer on the surface to separate. Depending upon the strength of the shock, the separated flow region can be small or large. In any case, it affects the properties of the reflected shock wave as well as the area available for the main flow. In addition, there is usually an expansion fan centered on the crown of the separation bubble and an oblique shock (called the reattachment shock) at the point of reattachment of the boundary layer.
- Growth of the boundary layer leads to a reduction in the cross-sectional area available for the main flow. This can eventually lead to choking and a low frequency instability condition known as buzzing under extreme sub-critical operation (Fig. 8.5). The pulsating or oscillating flow field associated with this instability can result in a flame blow-out in the combustor.

With these issues in mind, in practice, the boundary layer thickness is kept to the minimum possible by means of boundary layer bleeds. Here, the slow moving layers of fluid very close to the wall are sucked out through ports in the surfaces at regular intervals. This strategy addresses both the issues mentioned above:

- Since the momentum of the fluid in the boundary layer is higher now, flow separation is delayed for a given adverse pressure gradient. Hence, flow separation due to shock impingement, if it occurs, is confined to a small region.
- Removal of the slow moving fluid near the wall keeps the boundary layer thickness constant, thereby delaying or even avoiding buzzing of the intake.

More details on the stability aspects of 2D intakes are available in the book *Intake Aerodynamics* by Seddon and Goldsmith.

8.2.6 *Axi-Symmetric (Conical) Intake*

In our development so far, we have tacitly assumed the flow to be 2D for the sake of simplicity. Although the 2D intake is used almost exclusively in supersonic aircrafts, conical intakes also find a place. This type of intake was used in the SR71 Blackbird and is currently used in missiles that employ ramjet propulsion. In a conical diffuser, the streamlines behind the shocks are curved and there can be significant radial and axial pressure gradients. Nevertheless, the concepts and issues that have been discussed in connection with 2D intakes are equally applicable to conical intakes. The conical intake is axi-symmetric in shape and features a centerbody with compression surfaces inside a cowl. The compression surfaces on the centerbody are the same as the ramp surfaces in a 2D intake. The cowl design is also the same. The conical intake, like its 2D counterpart, can also be designed to be a purely external compression intake or a mixed compression intake. The centerbody is usually movable in the axial direction, to allow the intake to be operated near the critical mode for a range of flight Mach numbers. More details on the design of conical intakes are available in *Mechanics and Thermodynamics of Propulsion* by Hill and Peterson.

8.2.7 Thermodynamics of a Ramjet Engine

8.2.7.1 Ideal Ramjet Engine

The thermodynamic processes undergone by the air as it goes through an ideal ramjet engine are shown in Fig. 8.7. In this figure, processes a-1 (intake), 1–2 (combustor) and 2-e (nozzle) are assumed to be isentropic. Also, heat addition in the combustor is assumed to be a constant pressure process (1–2) without any loss of stagnation pressure. In other words, all the processes have been assumed to take place without any irreversibilities. Hence, on the whole, we have assumed the engine to be ideal. Both the static state and the corresponding stagnation state are shown in Fig. 8.7. In contrast to the turbojet and the turbofan engine, here, there is no work interaction in any of the processes. The only change in the stagnation state is across the combustor due to heat addition. Since the stagnation pressure remains the same, $P_{0,e} = P_{0,a}$ and if the flow expands to the ambient pressure in the nozzle, i.e., $P_e = P_a$, then it follows from Eq. 2.24 that $M_e = M_a$ (assuming γ remains the same). Note, however, that the exit velocities will be different since the exit static temperature is different from T_a. In fact, since $T_{0,e} > T_{0,a}$ and $M_e = M_a$, it is easy to see that $T_e > T_a$ and hence $V_e > V_a$.

8.2.7.2 Actual Ramjet Engine

As discussed in the previous section, the compression of the air in the intake is usually accomplished by a combination of oblique, normal shocks and an appropriately shaped passage. All of these processes are irreversible and so there will be an increase in entropy. Furthermore, in contrast to the situation in turbojet and turbofan engines, the Mach number at entry to the combustor in the case of a ramjet, although subsonic, will be high enough for compressibility effects to be significant. Hence, the heat addition process (if we assume the combustor cross-sectional area to be constant) will follow the Rayleigh line, and there will be a loss of stagnation pressure (as explained in the discussion following Eq. 2.26). The processes undergone by the air in an actual ramjet thus appear as shown in Fig. 8.8. Note that $P_{0a} > P_{0,1} > P_{0,2} > P_{0,e}$. Process

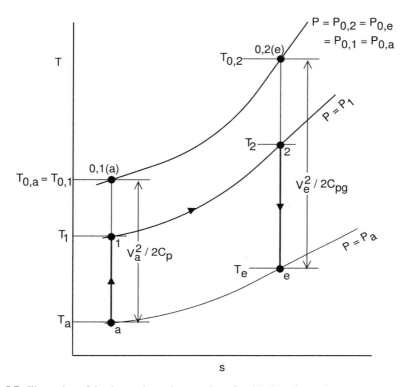

Fig. 8.7 Illustration of the thermodynamic operation of an ideal ramjet engine

a-1 has been shown in this figure as a single straight line, although in reality it would be a series of line segments each one representing an oblique or normal shock. The stagnation pressure recovery across the intake, $\eta_r = P_{01}/P_{0a}$ can be calculated from the military specification MIL-E-5008B as[4]

$$\begin{aligned}
\eta_r &= 1, \quad M_a \leq 1 \\
&= 1 - 0.075(M_a - 1)^{1.35}, \quad 1 < M_a < 5 \\
&= \frac{800}{M_a^4 + 935}, \quad M_a > 5.
\end{aligned} \tag{8.2}$$

The stagnation pressure loss across the combustor due to heat addition can be calculated using the Rayleigh table, if the combustor inlet Mach number and the amount of

---

[4] In fact, for a given flight Mach number M_a, this expression can be used to roughly configure the intake as follows. If we assume a sequence of shocks of equal strength terminating in a normal shock in the intake, then the stagnation pressure recovery across such a shock system can be calculated. This can be compared with the value given by Eq. 8.2 to determine the number of shocks required and hence the intake configuration.

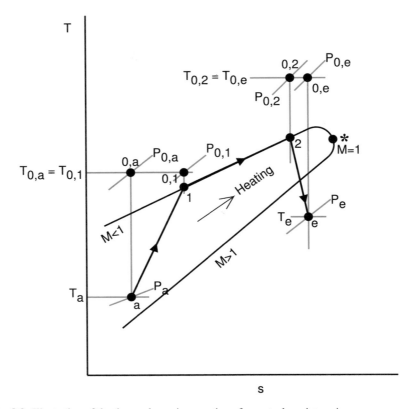

Fig. 8.8 Illustration of the thermodynamic operation of an actual ramjet engine

heat added are known. However, losses due to the flame holder cannot be determined exactly and can only be estimated. For the nozzle, the isentropic efficiency can be used to take into account irreversibilities during the expansion process.

8.2.8   Thrust Calculation

The expression for thrust and TSFC for a ramjet engine are the same as those for a turbojet engine, viz.,

$$\frac{T}{\dot{m}} = V_e - V_a + (P_e - P_a)\frac{R_g T_e}{P_e V_e} \tag{8.3}$$

$$TSFC = \frac{\dot{m}_f}{T} = \frac{\dot{m}_f/\dot{m}}{T/\dot{m}}. \tag{8.4}$$

The thrust calculation procedure is illustrated next using an example.

Example 8.1 A ramjet is designed for flight conditions of 12 kPa, 217 K and 738 m/s. The maximum temperature reached in the combustor is 2100 K. Assume $\eta_{nozzle} = 0.97$, $\eta_{comb} = 0.8$ and $\Delta P_{0,comb} = 8\%$. For air, take $C_p = 1.005$ kJ/kg K and $\gamma = 1.4$. For the combustion gases, take $C_{pg} = 1.148$ kJ/kg K and $\gamma_g = 1.333$, $H_{fuel} = 45,000$ kJ/kg.

Determine the specific thrust and the TSFC when the nozzle is (a) convergent and (b) convergent divergent designed for expansion to the ambient pressure.

Solution: Given $P_a = 12$ kPa, $T_a = 217$ K and $V_a = 738$ m/s. The flight Mach number

$$M_a = \frac{738}{\sqrt{0.4 \times 1005 \times 217}} = 2.5.$$

The freestream stagnation pressure

$$P_{0,a} = P_a \left(1 + \frac{\gamma - 1}{2} M_a^2 \right)^{\frac{\gamma}{\gamma-1}} = 205.03 \text{ kPa}$$

and stagnation temperature

$$T_{0,a} = T_a \left(1 + \frac{\gamma - 1}{2} M_a^2 \right) = 488.25 \text{ K}.$$

Upon using Eq. 8.2, we get $P_{0,1}/P_{0,a} = 0.87$. Thus, $P_{0,1} = 178.3761$ kPa. Also, $T_{01} = T_{0a} = 488.25$ K.

Since $\Delta P_{0,comb} = 8\%$, $P_{0,2} = (1 - 0.08)(178.3761) = 164.1$ kPa and $T_{0,2} = 2100$ K.

For the convergent nozzle,

$$P_{e,crit} = P_{0,2} \left(1 - \frac{1}{\eta_{nozzle}} \frac{\gamma_g - 1}{\gamma_g + 1} \right)^{\frac{\gamma_g}{\gamma_g - 1}} = 86.86 \text{ kPa}.$$

Since $P_{e,crit} > P_a$, the nozzle is choked. Hence

$$P_e = 86.86 \text{ kPa}$$

and

$$T_e = \frac{2}{\gamma_g + 1} T_{0,2} = 1802.6 \text{ K}$$

and

$$V_e = \sqrt{2C_{pg}(T_{0,e} - T_e)} = 826.37 \text{ m/s}.$$

Hence, the specific thrust can be evaluated as

$$\frac{T}{\dot{m}} = V_e - V_a + (P_e - P_a)\frac{R_g T_e}{P_e V_e} = 88.37 + 535.5 = 623.87 \text{ Ns/kg.}$$

Note that the pressure thrust is 6 times more than the momentum thrust owing to the fact that a convergent nozzle is used. To evaluate thrust specific fuel consumption, we start by applying the energy equation 2.17 to the combustor,

$$\dot{m}C_p T_{0,1} = (\dot{m} + \dot{m}_f)C_{pg} T_{0,2} - \dot{m}_f H_{\text{fuel}}.$$

Upon rearranging this, we get

$$\frac{\dot{m}_f}{\dot{m}} = \frac{(C_{pg} T_{0,2} - C_p T_{0,1})}{H_{\text{fuel}} - C_{pg} T_{0,2}} = 0.045.$$

This is the ideal fuel mass flow rate. The actual mass flow rate is given by $0.045/\eta_{\text{comb}} = 0.05625$. Therefore

$$TSFC = \frac{\dot{m}_f/\dot{m}}{T/\dot{m}} = \frac{0.05625}{623.87} = 0.325 \text{ kg/Nh.}$$

If the nozzle is a convergent divergent one designed for expansion to ambient pressure, then $P_e = P_a = 12$ kPa. The velocity of the fluid at the nozzle exit, V_e has to be determined. With reference to Fig. 7.9, since states es and 02 lie on the same isentrope,

$$\frac{T_{0,2}}{T_{es}} = \left(\frac{P_{0,2}}{P_a}\right)^{\frac{\gamma_g - 1}{\gamma_g}}$$

where we have used the fact that $P_{es} = P_e = P_a$, as these states lie on the same isobar. Hence

$$V_{es}^2 = 2C_{pg}(T_{0,2} - T_{es}) = 2C_{pg} T_{0,2}\left(1 - \frac{T_{es}}{T_{0,2}}\right).$$

From the definition of the nozzle isentropic efficiency, Eq. 7.15, the exit velocity can be calculated as

$$V_e = \sqrt{\eta_{\text{nozzle}} V_{es}^2} = \sqrt{2\eta_{\text{nozzle}} C_{pg} T_{0,2}\left(1 - \frac{T_{es}}{T_{0,2}}\right)} = 1494 \text{ m/s.}$$

Hence, the specific thrust can be evaluated as

$$\frac{T}{\dot{m}} = V_e - V_a = 756 \text{ Ns/kg.}$$

Note the increase in specific thrust (which is now entirely momentum thrust) due to expansion to ambient pressure in the convergent divergent nozzle. Although the pressure thrust is now zero, the increase in the momentum thrust more than offsets the decrease in the pressure thrust.

TSFC for this case can be calculated as

$$\text{TSFC} = \frac{\dot{m}_f/\dot{m}}{T/\dot{m}} = \frac{0.05625}{756} = 0.268 \, \text{kg/Nh}.$$

8.3 Turboramjet Engine

The turboramjet engine is a hybrid engine which uses a combination of an after-burning turbojet engine and a ramjet engine. The SR-71 Blackbird and the Concorde used this type of engine to propel the vehicle at supersonic cruise speeds. The engine is essentially an afterburning turbojet mounted inside a bypass duct as shown in Fig. 8.9. The engine operates in turbojet mode during takeoff, landing and subsonic flight speeds. The afterburner is turned on to accelerate through the sonic barrier and reach supersonic flight speeds. Once supersonic cruise speed is attained, the engine operates in the ramjet mode. The mode of operation of the engine is controlled by using bypass flaps located just downstream of the intake. During subsonic and tran-

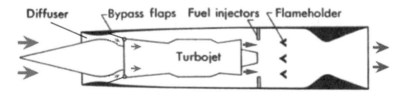

a) Bypass flaps allow flow into turbojet

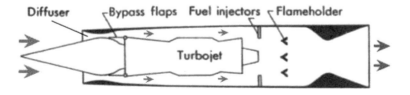

b) Bypass flaps block flow into turbojet
during ramjet mode

Fig. 8.9 Schematic of a turboramjet engine. *Source* http://www.aerospaceweb.org/question/propulsion/q0175.shtml

sonic flight speeds, these controllable flaps close the bypass duct and force the air directly into the compressor section of the turbojet. During supersonic cruise, the flaps block the flow into the turbojet, and the engine operates like a ramjet using the afterburner to produce thrust. This design approach allows the vehicle to take-off and land using normal runways with the turbojet, and cruise at supersonic speeds using the ramjet engine, by combining the best features of both the turbojet and ramjet into a single engine. The intake for the turboramjet engine is more complex in design and is usually a variable geometry intake, since it has to operate efficiently from subsonic through transonic to supersonic flight speeds.

8.4   Scramjet Engine

Turbojet engine and its derivatives cannot be used for hypersonic Mach numbers because the propulsive efficiency is so low as to be unviable for operation. Similarly, ramjets are effective only up to Mach 4. In a ramjet engine, combustion takes place at subsonic speeds. So, the incoming air is decelerated through a series of oblique shocks terminating in a normal shock. The flow at the exit of the intake thus becomes subsonic. As the cruise Mach number increases, the pressure and the temperature of the air entering the combustor also increases. When the flight Mach number attains values higher than 4, the temperature of the air entering the combustor is so high that addition of any more heat in the combustion chamber usually results in molecular dissociation. Once dissociation starts, conversion of enthalpy to velocity and hence thrust, becomes highly inefficient. The total pressure loss is also drastic across the normal shock making the operation of ramjet difficult at hypersonic Mach numbers. So to propel the vehicle at hypersonic speeds the air should still be moving at supersonic speeds when it enters the combustor. Thus, the combustion process itself takes place when the flow is supersonic. Hence, the name, Supersonic Combustion Ram jet engine.

8.4.1   Operation of Scramjet Engine

In a scramjet engine, the air stream coming at hypersonic Mach numbers is compressed by a serious of oblique shocks as shown in Fig. 8.10. The compressed air stream enters the combustor typically at Mach 2–2.5. The fuel is mixed with the supersonic air stream and burnt inside the combustor. The hot supersonic flow is allowed to expand in the nozzle, thereby generating the thrust required for propelling the vehicle at hypersonic speeds. Intake design is very important as it is necessary to decelerate and compress the air with as little loss of total pressure as possible. Combustor design is also critical since the combustion must take place at supersonic speeds and in such a way that enough heat is added to generate the required thrust. Too much heat addition will result in a normal shock standing at the combustor

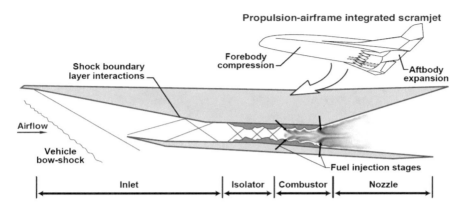

Fig. 8.10 Schematic illustration of a scramjet engine. *Source* http://www.nasa.gov/centers/dryden/news/FactSheets

entrance. This is known as an inlet interaction. On the other hand, if too little heat is released, then combustion cannot be sustained, resulting in a blowout. The velocity of the air as it enters the combustor is usually of the order of 1 km/s and the length of the combustor itself is of the order of 1 m or so. Thus, the total residence time of the flow in the combustion chamber is about 1–2 ms. Within this short duration, the fuel should be mixed with the air and burnt in the combustor and this requires highly efficient mixing strategies and effective flame stabilization methods.

Since the main stream flow velocity is high, the injected fuel (regardless of the direction of injection) tends to get blown downstream and does not travel any appreciable distance in the lateral or the vertical direction at all. In other words, it does not mix with the air. Mixing can be improved by cross-flow. However, any aggressive mixing strategy can lead to the formation of a normal shock at the entry to the combustor, which is undesirable. For the same reason, mechanical flame stabilizers (such as V-gutters) cannot be used in the supersonic cross-flow. Therefore, flame stabilization has to be achieved through aerodynamic features in the combustor design. From the point of view of thrust generation, combustor entry Mach numbers should be as high as possible, so that more heat can be added without thermally choking the flow. From a flame stabilization perspective, the opposite is desirable. Realization of a working scramjet engine continues to be a challenge in the face of such conflicting and difficult requirements.

Suggested Reading

1. Babu V (2015) Fundamentals of gas dynamics, 2nd edn. Athena Academic
2. Hill PG, Peterson CR (1991) Mechanics and thermodynamics of propulsion, 3rd edn. Addison Wesley
3. Mattingly JD (1996) Elements of gas turbine propulsion. Aeronautical and aerospace engineering. McGraw-Hill
4. Oates GC (1997) Aerothermodynamics of gas turbine and rocket propulsion, 3rd edn. AIAA education series
5. Oates GC (1985) Aerothermodynamics of aircraft engine components. AIAA education series
6. Saravanamuttoo HIH, Rogers GFC, Cohen H (2004) Gas turbine theory, 5th edn. Pearson Education
7. Shepherd DG (1956) Principles of turbomachinery. Macmillan Publishing Co
8. Turns S (2000) An introduction to combustion, 2nd edn. McGraw-Hill International Editions

Index

© The Author(s), under exclusive license to Springer Nature Switzerland AG 2022
V. Babu, *Fundamentals of Propulsion*,
https://doi.org/10.1007/978-3-030-79945-8

Printed in the United States
by Baker & Taylor Publisher Services